# Integrating Racial Justice Into Your High-School Biology Classroom

In this guide, educators and authors David Upegui and David E. Fastovsky offer a pedagogical prescription for how you can integrate the study of racial justice with evolutionary biology in your existing high-school biology curriculum.

Designed as a practical manual for teaching, the chapters focus on teaching concepts of equity through evolutionary biology modules, a cornerstone for building students' scientific understanding of biotic diversity. The book provides pedagogical components alongside historical and scientific components, with contextual chapters that give teachers the background knowledge to understand the historical relationship between science and racism for topics such as natural selection, social justice, and American slavery and colonization. Ready-to-use lesson plans are situated in a historical and theoretical context of science as it relates to racial oppression, and demonstrate how rigorous science education can lead to your students' liberation and personal empowerment despite the historically problematic history of some applications of science. These lesson plans and classroom exercises are presented in a way that introduces the timely extra dimension of anti-racism into existing biology curricula without significantly increasing teaching loads. The contextual material provided allows the lessons to be implemented across a variety of classrooms regardless of initial familiarity with DEI.

Ideal for secondary biology teachers and their students, particularly in grades 10–12, this book synthesizes timely ideas for high-school educators, harnessing the power of rigorous science to combat marginalization. Lessons and activities have been classroom-tested and are aligned with three different standards: Next Generation Science Standards (NGSS); College board (AP Biology); Vision and Change; and use the 5E format.

**David Upegui** is a Latino educator, with 12 years of teaching biology in an inner-city US high school. He has won the Presidential Award for Excellence in Math and Science Teaching and the Evolution Education Award [NABT].

**David E. Fastovsky** is Professor Emeritus of Geosciences at the University of Rhode Island, United States, where, for 36 years, he taught evolutionary biology and Earth and biotic history.

## Also Available from Routledge Eye On Education

**www.routledge.com/k-12**

**Facilitating Conversations about Race in the Classroom**
By Danielle Stewart, Martha Caldwell and Dietra Hawkins

**When Black Students Excel**
**How Schools Can Engage and Empower Black Students**
By Joseph F. Johnson Jr., Cynthia L. Uline, and Stanley J. Munroe Jr.

**The Antiracist World Language Classroom**
By Krishauna Hines-Gaither and Cécile Accilien

**Teaching as Protest**
**Emancipating Classrooms Through Racial Consciousness**
By Robert S. Harvey and Susan Gonzowitz

# Integrating Racial Justice Into Your High-School Biology Classroom

## Using Evolution to Understand Diversity

David Upegui and David E. Fastovsky

Routledge
Taylor & Francis Group
NEW YORK AND LONDON

Designed cover image: © Getty Images

First published 2024
by Routledge
605 Third Avenue, New York, NY 10158

and by Routledge
4 Park Square, Milton Park, Abingdon, Oxon, OX14 4RN

*Routledge is an imprint of the Taylor & Francis Group, an informa business*



*Library of Congress Cataloging-in-Publication Data*
Names: Upegui, David, author. | Fastovsky, David E., author.
Title: Integrating racial justice into your high-school biology classroom : using evolution to understand diversity / David Upegui, David E. Fastovsky.
Description: New York, NY : Routledge, 2024. | Includes bibliographical references.
Identifiers: LCCN 2023014606 (print) | LCCN 2023014607 (ebook) | ISBN 9781032523842 (hardback) | ISBN 9781032529141 (paperback) | ISBN 9781003409144 (ebook)
Subjects: LCSH: Biology—Study and teaching (Secondary) | Human evolution—Study and teaching (Secondary) | Racial justice in education.
Classification: LCC QH315 .U64 2024 (print) | LCC QH315 (ebook) | DDC 570.71—dc23/eng/20230501
LC record available at https://lccn.loc.gov/2023014606
LC ebook record available at https://lccn.loc.gov/2023014607

ISBN: 978-1-032-52384-2 (hbk)
ISBN: 978-1-032-52914-1 (pbk)
ISBN: 978-1-003-40914-4 (ebk)

DOI: 10.4324/9781003409144

Typeset in Palatino
by Apex CoVantage, LLC

# Contents

# About the Authors

**David Upegui** is a Latino immigrant who found his way out of poverty through science. He currently serves as a science teacher at his alma mater, Central Falls High School (RI) where he has taught biology for the past 12 years, and as an Adjunct Professor of Education at Brown University. He received the NABT's Outstanding Biology Teacher Award (2021) and the Evolution Education Award (2014) as well as the Presidential Award for Excellence in Mathematics and Science Teaching in 2019. Upegui started, and runs, the school's Science Olympiad team and has written extensively on science education and appropriate pedagogy. He completed his doctoral degree in education at the University of Rhode Island, focusing on science education and social justice. Reach Upegui on Twitter (@upeguijara)

**David E. Fastovsky** is a paleontologist and Earth historian at the University of Rhode Island, where he was a Professor for 36 years and co-developed the university's first introductory course in evolutionary biology. He is a two-time Fulbright Scholarship recipient, a Fellow of both the Geological Society of America and the Paleontological Society, and the author of well over 100 scientific publications, including a successful textbook on the evolution of dinosaurs, currently in its fourth edition. He served as Dr. Upegui's doctoral advisor.

# Acknowledgements

This book had many contributors who helped shape the ideas and focus the writing. Initial guidance was provided by Upegui's University of Rhode Island doctoral committee members (Drs. Julie Coiro, Stefan Battle, and Rudolf Kraus) to whom we are grateful for their time, guidance, and support.

Several people provided much appreciated direct input and guidance. Such is the case with Dr. Kirstin Milks (Bloomington [Indiana] HS South) who read and commented extensively on the manuscript, *really* encouraged us, and made key suggestions for the text. Dr. Agustin Fuentes (Princeton University) reviewed the lessons and provided thoughtful comments. Ms. Tracy Lafreniere (RI Teacher of the Year, 2016) also provided feedback on the readability of the manuscript. Dr. Jon Marks (University of North Carolina, Charlotte) reviewed potential student-answers to questions in the lessons; Dr. Joseph Graves Jr. (North Carolina Agricultural and Technical State University, Greensboro) provided practical and inspirational ideas, as well as a generous open door for consultation. Drs. Fuentes, Graves, and Marks also recorded wonderful YouTube videos to accompany each of the lessons.

Special thanks are due to Julia Dolinger and Sophie Ganesh, our Editors at Routledge, who early on expressed an interest in our work, and have actively moved it forward as it developed. Under their care, six professional secondary-school educators reviewed and commented extensively on this manuscript; the combined efforts of all of these individuals improved this work substantially.

The students from Central Falls are truly amazing humans in dire need of educational opportunities to empower them so they can fulfill their promise, not to say their destinies, and make their world (and ours) a better place. We are forever grateful for their openness, tenacity, and talent—only a small fraction of which could we showcase in the stories throughout this book.

Finally, this work would not have come to fruition without the support of our wonderful wives, Maria (Upegui's) and Lesley (Fastovsky's) and our children—Isaac and Nicolas (Upegui's) and Naomi and Marieke (Fastovsky's)—who are so much more than either of us ever imagined and are surely the lights of our lives. They have made us more patient, more grateful, more kind, more open-minded, and more joyous.

It is our hope that with this book teachers feel a push to see what is possible and then make it happen.

# Preface

This book is an extended prescription for the inclusion of social justice in high-school biology curricula. In it, the non-intuitive relationship between evolutionary biology, a science discipline, and racial justice, a branch of the humanities, is elaborated both academically and practically. The book develops the premise that this relationship is best explored at the high-school level and, most importantly, provides a template for how this can be done. Naturally, a single lesson, or even a well-developed set of lessons is not enough to melt away 400 years of societal oppression, but as a meaningful step towards "leveling the playing field," we offer here a prescription for an evolutionary biology curriculum module that, put directly, is intended to be no less than a "catalyst for academic achievement and social transformation" (Morales-Doyle, 2017, p. 1034).

Chapters 2–4 constitute the focal point of this work: the full prescription for the integration of racial justice into biological science curricula. They consist of a suite of practical activities for students, as well as a broad array of associated resources, set in a theoretical and historical context. While Chapters 2–4 constitute the lessons and activities themselves, the theoretical underpinning and rationale for these, both pedagogic and scientific, are developed in Chapters 5–7. Finally, Chapters 8–10, are the supporting players; they constitute the historical context for the lessons, and explore both the historical and philosophical bases of the relationship between biological science and "race."

*Therefore, with perhaps the exception of Chapter 1, this book is not meant to be read sequentially as one might read, for example, a novel. Rather, it is meant to offer lessons for high-school biology students, as well as provide significant and meaningful context for both students and instructors.* Users should go to the chapters as resources and integrate the information contained in them as needed. We provide neither the first nor the last words on any of the subjects treated here; but we provide sufficient context that with our book in hand, teachers can teach—and learners can learn—these lessons with a degree of confidence that they have been exposed to most of the key ideas associated with American racial history and its support—or lack of it—in the biological sciences.

For one of us (Upegui), personal history as a student of color and, later, as a science teacher in Central Falls High School, the most economically disadvantaged school in the state of Rhode Island, provided the impetus for this work. Stories from Upegui's life bracket each of the chapters, and are included to position readers to better understand how science-based racism and its byproducts can deeply influence and damage young minds. Its opposite—robust science in the service of understanding exactly what races are—and are not—empowers them.

Imbued with Enlightenment-originated logical systems that poorly accommodate mutually exclusive ideas, westerners generally, and western students, in particular, have difficulty in acknowledging the truth in two apparently antithetical perspectives. Yet historically we paired slavery with freedom; colonization with liberation. Indeed, the United States has carried out astounding acts of altruism (the Marshall Plan comes to mind), but it also systematically oppressed peoples, both at home and abroad, in colonies (Immerwahr, 2019), despite the "truths" that, in its most seminal documents, it holds to be "self-evident." In other cultures and times, embracing contradictions and paradoxes has been regarded as increased wisdom. F. Scott Fitzgerald concurred: "The test of a first-rate intelligence is the ability to hold two opposed ideas in the mind at the same time, and still retain the ability to function."

In science the need to hold two different truths at the same time has emerged in several areas. For example, in the wave-particle duality concept in quantum mechanics predicts that electrons will behave as waves and as particles. Similarly, in biological evolution both a gradual, continuous evolution (phyletic gradualism), and intervals of morphological stability followed by abrupt evolutionary change (punctuated equilibrium) generate patterns of biotic diversity.

It is in this spirit of dualities that this work is presented. Science has served as both a tool of empowerment and a tool of oppression. The time has come, therefore, to face the history that has created our current problems, teach it, and show our students that there is another, better way to see themselves and their inheritance: America's future.

Although "hard" science is commonly caricatured as impartial and somehow beyond the subjective passions that inflame the "soft" sciences, it too has a history of interpretations that have not always been impartial or even logical, but instead were driven by passions and emotion. How could it be otherwise? Science is a human endeavor, carried out by passionate and emotional humans. Yet, through science education we should and can empower students to envision and create a better society. The underlying premise of this book, therefore, is that science education, properly situated in its historical context, can make the world a better place.

Biology is surely the appropriate subject in which to address racism, because throughout its history, racism has always found its ultimate justification in the assertion of *biological* "truths" (most notoriously, that of white racial superiority). Evolutionary biology is the proper vehicle for considering racism, because evolutionary biology is ultimately the study of the origin and processes of biotic diversity, which is of course exactly what races are presumed to represent. These subjects are best introduced at the high-school level, because it is at this time that students are able to understand and synthesize the history and the science fully, and yet are not so fixed in their opinions and world views that they cannot absorb the societal significance of the lessons they've learned.

But to be clear: this book is a *not* a guilt-drenched restatement of white despicability and specious science. Although it comes to grips with a fraught, even, at times, sordid past, it is really about the future: the future of an ever more diverse American society and what life in it could look like. Perpetuating past mistakes and delusions won't take us where we need to go; however, sharing with the next generation a clear-eyed look at who we have been and who we are, where we want to be, and how we might get there, seems to us, as educators, the most important thing that we can do.

# Part I
## Introduction

# 1

# Why Introduce Racial Justice in a Biology Class?

*When Michael Rodriguez died in January of 1999, we were devastated. He was the coolest and most gregarious person in our crew. I had known Mikey since High School and once I was studying biology in college, along with Roger, we shared an apartment on Hunt street, Central Falls. Mikey defended my nerdiness with his usual swag, "Upegui, may be missing out in the parties now, but eventually, he will be able to do whatever he wants." Growing up in Central Falls meant that we knew poverty and struggles. His untimely death at the tender age of 23, left me and rest of his friends and family unsettled. Unable to provide any assistance to him while he struggled, I grew angry and confused. Seeking a way to show both my disillusionment and frustration with his death, I sought ways to celebrate his life and remember his contributions to our lives. What came to mind was a tattoo! But it had to be something special that showed the connection between Mikey and me, as well as my newly found respect for science.*

*With tears in my eyes as I remembered Mikey, I described to the tattoo artist what I was feeling and how this tattoo would represent both a loss and a connection. He asked me to show him some visual ideas. I took out my introductory Biology textbook and opened up to the page showing anaphase—the stage of mitosis when the newly-copied chromosomes move to opposite poles of the cell getting ready to split and become two independent cells. I thought this was a deep and powerful idea, as it included the concept of togetherness, and separation. The artist liked the idea until he saw a diagram of anaphase. His response was, "that is a peanut, I don't care what you think it looks like. To everyone else who doesn't know biology, that's a freaking peanut. And I ain't doing a freaking peanut on your arm—I will lose customers." Once he pointed this out, I agreed and moved on to my second idea—DNA.*

DOI: 10.4324/9781003409144-2

*Deoxyribose Nucleic Acid, could there be a more visually poetic idea in biology? A double stranded helix that carries the genetic instructions for all known living organisms—the molecule that connects us to each other and to all living things, both past and present! Yes, it would be DNA on my right arm, and it would encompass both my respect for science and the memory of Mikey.*

*To my aunts and uncles in Colombia, tattoos were still a taboo and the mark of shady street characters. But I had a tattoo and yet they loved me. Therefore, instead of immediately stressing their disapproval of my tattoo, my aunts asked me what it meant. I began by telling them that this was a symbol that represented my friend who had taken his life and my connection to him. Immediately, my aunt Alba jumped up and said, "of course, it is the snake crawling up the arm to symbolize loss." To which my aunt Celly said, "no, it is a ladder that is climbing to heaven towards his friend." My aunt Dora claimed that she saw a "rope that connected people," and my aunt Olga suggested that it was clearly a "scientific tool." The discussion went on for a while with each of the aunts providing possible explanations to the tattoo.*

*After a few minutes, they let me speak and I told them that DNA is the molecule that lives in our cells and in the cells of all living things. It is in fact that way that our bodies store information about how to develop, function, and reproduce. To me, this tattoo was a way to show both love for Mikey, with a clear appreciation for the knowledge I gained through science. Regardless of my explanation, after a few seconds of silence, they all agreed that my tattoo was a ladder reaching for heaven. Science was empowering to me as it allowed me to see the world differently; but more significantly, science intimately connected very foundations of the United States and modern civilization. If only I could convey this to my students!*

## Science and Society

The empowerment of Black, Indigenous, People of Color (BIPOC) as a collective group is essential because (a) an extraordinary number of the problems and issues that we face as a society arise from BIPOC[1] oppression, and (b) because the world can only benefit from diverse thinking and solutions. There is ample empirical evidence for the kind of society and environment produced by the domination of racist and elitist approaches. We have been left with a short-sighted legacy of environmental devastation, poverty, and disenfranchisement, which only single out a select few. It is obvious that there is a need for us to find better ways to empower the next generation, so that it, at least, has a chance to develop long-term, sustainable, and equitable solutions to societal and environmental problems.

Members with power in society have long justified the differences in living conditions and outcomes between different races based on what turn out to be non-existent, intrinsic biological differences. This removed them of all responsibility to repair that which was inequitable. But does the science really lead to this conclusion, or has the science been badly misunderstood? The answer to that question, and the underlying premise of the lessons that we detail in this book, is an unambiguous "no." The science is clear and historically, individuals with non-scientific agendas have clearly misconstrued it. The emptiness of their arguments is laid bare in this book.

## Science Education and Racial Justice

Our view is that science education has a vital role in empowering people. Lewontin (1991) explicitly highlighted the role of science and science education in the formation of societal relationships:

> *Science . . . has a function in the formation of consciousness about the political and social world. Science in that sense is part of the general process of education, and the assertions of scientists are the basis for a great deal of the enterprise of forming consciousness. Education in general, and scientific education in particular is meant not only to make us competent to manipulate the world but also to form our social attitudes.*
>
> (pp. 103–104)

Freire (1970) concurs, and states that the goal of education is for students to become agents for social justice, challenging inequities and the political status quo (see also Giroux, 2004). Morales-Doyle (2015) prescribes a theoretical framework in which students apply the logic of scientific thinking to identify the unfairness of social structures. It makes logical sense that we need to provide context around every single idea we learn because through this, students gain real understanding, and thus power, in society.

Science can also be used to liberate society from sociologically as well as scientifically unsupported ideas. Hodson (2009) eloquently articulated the necessity for the scientific literacy, of our multicultural public in the following passage:

> *In short, why does it matter what image of science is presented and assimilated [in schools]? It matters insofar as it influences career choice, and so may have long-term consequences for individuals. It matters if the curriculum image of science is such that it dissuades creative, non-conformist and politically conscious individuals from choosing to pursue science at an advanced level. It matters if the image of science is such that it dissuades women, members of visual minority groups and students from lower socioeconomic status homes from emerging science-related careers or seeking access to higher education in science and engineering because they do not see themselves included and*

> *represented in the science curriculum. It matters if our politicians, public servants and industrialists are so ignorant of scientific and technological issues that their decision making is ill-informed and uncritical. . . . Failing to provide every student with an adequate understanding of the nature of science runs counter to the demand for an educative citizenry capable of responsible and active participation in a democratic society.*
>
> (p. 142–143)

Thankfully, a well-constructed science curriculum not only dismisses erroneous ideas using scientific evidence, but it also enables liberation and empowerment, because *it turns out that the same tools needed for good science are also those needed to explore and change unjust social constructs*. Such tools include asking questions, planning and carrying out investigations, analyzing and interpreting data, engaging in argument from evidence, and obtaining, evaluating, and communicating information (National Research Council, 2012)

## High School Biology Education and Racial Justice

Scientific theories are superseded when new evidence and ideas demonstrate that the previous explanations are inadequate, incomplete, or wrong (see Kuhn, 1970). Such is the case for phrenology, spontaneous generation, evolution by the inheritance of acquired characters, triple-stranded DNA, and countless other less-notorious abandoned scientific hypotheses. However, the hypothesis that races reflect innate, immutable biological differences has been notably resistant to rejection. Its persistence surely reflects and a deep-seated, irrational emotional response to "the other," and perhaps even more importantly, its utility in maintaining political power structures, rather than scientific merit. Understanding what race actually is, therefore, provides an *entré* to understanding the social world as well as its natural counterpart. (Graves & Goodman, 2021).

Biology is one of the core science courses taught in virtually all US high schools. For most of these students, in the midst of animal dissections, food webs, the stages of mitosis, and indeed, a well-packed biology curriculum, there may appear to be little room for an exposition of racial justice. Yet, perhaps the most important opportunity for students to connect racial justice with a history of oppression based on skin pigmentation and other superficial features, comes in biology classes during units in which evolutionary theory is explored. The hypothesis of evolution by natural selection (Chapter 7) is the scientific explanation of the origin of species, and for the unity and the diversity of life on Earth (Kampourakis, 2014). Taught and learned correctly, evolutionary biology has the potential to teach students about meaningful biotic connections, and to help them apply their insights beyond biology. In its absence, biology is a jumble of disaggregated facts that resist memorization, let alone understanding; in its presence, it is a cohesive intellectual construct (Dobzhansky, 1973).

The importance of the topic of evolution in a biology curriculum is highlighted in the "Life Science strands" of the Next Generation Science Standards (NGSS Lead States, 2013) since it makes up one of the five "topics" (National Research Council, 2012) and for AP bio curricula it is one out of the four "Big Ideas" (College Board, 2019). Therefore, there can be no doubt that evolution must be presented and assessed as part of biology class.

Evolutionary biology has direct topical bearing on social justice. Its significance here is that it is the appropriate and the ideal tool with which to trace the origin and diversification of species (including humans) on Earth, as well as the appropriate and ideal place for introducing the appearance of "races" as localized responses to ephemeral environmental conditions, rather than genetically-based, deep-seated differences among biologically dissimilar subspecies of hominins, derived from different ancestors. It is in evolutionary biology in which the biological significance of races—there is none—can be ascertained and, more importantly for our purposes, be directly taught.

Finally, there are important developmental reasons for teaching social justice to high school students. The physical changes in gray matter, white matter, and neurotransmitters from late childhood to early adulthood are dramatic and have implications for understanding emergent modalities of sophisticated thinking. The changes are both progressive (building up) and regressive (reducing and fine-tuning) and are sensitive to maturational and environmental influences (Jetha & Segalowitz, 2012). The temporal lobes which are associated with language, emotion, and memory, have a peak of cortex development around 16.7 years for males and 16.2 years for females—high school age (Jetha & Segalowitz, 2012). The thickening of this lobe is linked with emotional understanding, which involves the awareness of self and others and activates frontal and parietal/temporal regions (Dahl, 2004).

During adolescence, skills increase in complexity as young people gain more independence from their parents and spend more time with peers by taking on new social roles and responsibilities (Vera-Estay et al., 2015). Adolescents have amplified opportunities for autonomous decision-making, including decisions about complex moral issues that may have lifelong individual and societal consequences (Adams & Berzonsky, 2006). Some moral conflicts in social contexts warrant conscious evaluation, particularly when dilemmas are complex, the consequences are serious, or competing values and interests are at play. This is a crucial component in the moral decision-making process which requires the thinking mechanism through which moral judgments about right and wrong are attained.

Chiasson et al. (2017) demonstrated that there is a linear increase in moral reasoning from childhood to late adolescence with significant development culminating between the ages of 15 to 17. Additionally, evidence from social neuroscience research shows that brain changes occurring during adolescence are associated with the development of social cognition and executive functioning and the improvement of moral reasoning skills (Blakemore, 2008; Blakemore & Choudhury, 2006; Casey et al., 2005; Paus, 2005). From a theoretical point of view, adolescence is a critical phase for the development of abstract thinking skills, leading to a fuller integration of moral principles and values that are incorporated into the self-concept (Hardy & Carlo, 2011). Therefore, it is not until this developmental milestone that adolescents acquire the increasingly stable and mature organization of the identity system (Fischer & Lamborn, 1989) required to comprehend both the comprehensive use for science and the impetus for social justice, called for here.

## Conclusion

Evolutionary biology and social justice work well together in high-school curricula because:

(a) Evolutionary theory has an explicit biological relationship to social justice because the facts and logic of evolution strongly argue for the key tenets of social justice;
(b) Given a long history in which race-based "science" was carried out to make a political or social point (i.e., the hypothesis under study was not actually being tested with an eye towards its falsification), teachers have a unique and powerful opportunity to teach the difference between specious and misleading science, and precise and rigorous science;
(c) Historically, every instance of white supremacy asserted since the 1400s has always ultimately relied upon a diverse palette of biological explanations to justify a social hierarchy;
(d) High school is the first—and perhaps the most important—time when students are at an appropriate developmental stage to become aware of, to question, and to ultimately absorb concepts of social right and wrong (Blakemore, 2008, 2012; Paus, 2013; Koenis et al., 2018); as well as evolutionary theory (Kampourakis, 2013). The possibility that science can be both liberating and oppressing, requires

a maturity of thought that is first found in adolescents (Adams & Berzonsky, 2006); and

(e) The potential discordance between the logic of scientific thought and the belief systems of friends and family engenders a certain independence of thought in students and can contribute to their intellectual maturation, as they grapple with ideas that may diverge from what their intimates (and even they themselves) still believe.

The teaching of evolutionary theory leading to social justice works at the high-school level both in terms of science education and personal empowerment. No other science can make that claim, so it is evolutionary theory that is best suited to unite biology, to empower by science education, and to underpin racial justice. If we are to ask what is to be done to stop the vicious endless cycle of domination, subjugation, oppression and racism—our answer, then, is to begin to empower the next generation by teaching it these topics during high school. Therefore, this book advocates for and documents the relationships—scientific, social, and historical—between evolutionary biology and racial justice.

Our experiences have led us to firmly believe that rigorous classes have the potential of empowering students in emergent ways. Researchers have posited that teachers are the single most important classroom factor in a child's learning achievement (Hattie, 2017). During the dissertation proposal defense that would eventually mature into this book, a question was asked about the reason for Upegui's students' accomplishments: Is the post-graduation success of his students exclusively due to Upegui being a Latino teacher teaching mostly Latino children in his (heavily Latino) hometown (because, as it was said, "he is in his sauce")? This dismisses the role of the science content being taught. The reality, as is the case with most social endeavors, is that student success is a complex phenomenon that requires appropriate pedagogical approaches along with content material that engages students.

We have asserted in this Chapter that biology classes are uniquely positioned to become a place for liberation, where students learn how to utilize the tools of science to create a better, more just world. Even in a highly standardized curriculum, biology teachers still have meaningful ways to highlight important parts of the material and find approaches to cover the content that are productive for students. We are living in a moment in history that includes the most available information for the masses, and with this tremendous amount of information there is a need and responsibility to learn how to discern facts and utilize our collective knowledge for the betterment of humanity. Our interconnected global society faces grave problems including a hotter planet, political turmoil, overpopulation, wars, and social instability.

To solve the problems that, sadly, our students will inherit, they will require our concerted efforts for the sake of humanity, and we cannot think of a better place for this to begin than in biology classrooms.

*I have told the story of Mikey and my tattoo to all of my classes because they want to know why I have a tattoo of DNA. Since I began teaching, handfuls of students have gotten DNA tattoos based on mine. Even though the stigma around tattoos is not what it was twenty years ago, I have never promoted tattoos to my students—but on their own, they decided that a tattoo would allow them to acknowledge the role played by science in their new empowerment.*

## Note

1 The widely used acronym BIPOC first appeared in 2013 and is currently in common usage largely due to the inclusion of Indigenous people among those who have been historically oppressed (Garcia, 2020).

## References Cited

Adams, G.R., & Berzonsky, M.D. (2006). *Blackwell Handbook of Adolescence*. Oxford: Blackwell.

Blakemore, S.J. (2008). Development of the Social Brain During Adolescence. *Quarterly Journal of Experimental Psychology (Hove)*, *61*(1), 40–49. https://doi.org/10.1080/17470210701508715. PMID: 18038337.

Blakemore, S.J. (2012). Imaging Brain Development: The Adolescent Brain. *NeuroImage*, *61*(2), 397–406. https://doi.org/10.1016/j.neuroimage.2011.11.080

Blakemore, S.J., & Choudhury, S. (2006). Development of the Adolescent Brain: Implications for Executive Function and Social Cognition. *Journal of Child Psychology and Psychiatry*, *47*, 296–312.

Casey, B.J., Tottenham, N., Liston, C., & Durston, S. (2005). Imaging the Developing Brain: What Have We Learned About Cognitive Development? *TRENDS in Cognitive Sciences*, *9*, 104–110.

Chiasson, V., Vera-Estay, E., Lalonde, G., Dooley, J.J., & Beauchamp, M.H. (2017). Assessing Social Cognition: Age-Related Changes in Moral Reasoning in Childhood and Adolescence. *The Clinical Neuropsychologist*, *31*(3), 515–530. https://doi.org/10.1080/13854046.2016.1268650

College Board AP® BIOLOGY Course and Exam Description (Revised). (2019). https://apcentral.collegeboard.org/pdf/ap-biology-course-and-exam-description-0.pdf?course=ap-biology

Dahl, R.E. (2004). Adolescent Brain Development: A Period of Vulnerabilities and Opportunities. Keynote Address. *Annals of the New York Academy of Sciences, 1021,* 1–22. https://doi.org/10.1196/annals.1308.001. PMID: 15251869.

Dobzhansky, T. (1973). Nothing in Biology Makes Sense Except in the Light of Evolution. *American Biology Teacher, 35*(3), 125–129.

Fischer, K.W., & Lamborn, S.D. (1989). Mechanisms of Variation in Developmental Levels: Cognitive and Emotional Transitions during Adolescence. In A.K. Ribaupierre (Ed.), *Transition Mechanisms in Child Development: The Longitudinal Perspective,* pp. 33–67. Cambridge: Cambridge University Press.

Freire, P. (1970). *Pedagogy of the Oppressed.* New York: Continuum.

Garcia, S.E. (2020, June 17). Where Did BIPOC Come from? *New York Times.* www.nytimes.com/article/what-is-bipoc.html

Giroux, H.A. (2004). Public Pedagogy and the Politics of Neo-Liberalism: Making the Political More Pedagogical. *Policy Futures in Education,* 2(3–4), 494–503.

Graves, J.L., & Goodman, A. (2021). *Racism, Not Race: Everything You Wanted to Know About Race: But Were Afraid to Ask.* New York: Columbia University Press.

Hardy, S.A., & Carlo, G. (2011). Moral Identity: What is it, How Does it Develop, and is it Linked to Moral Action? *Child Development Perspectives,* 5(3), 212–218. https://doi.org/10.1111/j.1750-8606.2011.00189.x

Hattie, J. (2017). *Hattie Ranking: 252 Influences and Effect Sizes Related to Student Achievement.* https://visible-learning.org/backup-hattie-ranking-256-effects-2017/

Hodson, D. (2009). *Teaching and Learning About Science.* Rotterdam, The Netherlands: Sense Publishers.

Jetha, M.K., & Segalowitz, S.J. (2012). *Adolescent Brain Development: Implications for Behavior.* Waltham, MA: Elsevier.

Kampourakis, K. (2013). Teaching About Adaptation: Why Evolutionary History Matters. *Science & Education, 22,* 173–188. https://doi.org/10.1007/s11191-011-9363-2

Kampourakis, K. (2014). *Understanding Evolution.* Cambridge: Cambridge University Press.

Koenis, M.M.G., Brouwer, R.M., Swagerman, S.C., van Soelen, I.L.C., Boomsma, D.I., & Hulshoff Pol, H.E. (2018). Association between Structural Brain Network Efficiency and Intelligence Increases during Adolescence. *Human Brain Mapping, 39*(2), 822–836.

Kuhn, T.S. (1970). *The Structure of Scientific Revolutions.* Chicago: University of Chicago Press.

Lewontin, R.C. (1991). *Biology as Ideology.* New York: HarperCollins.

Morales-Doyle, D. (2015, October 21). *Science Education as a Catalyst for Social Change? Justice-Centered Pedagogy in Secondary Chemistry*. Doctoral dissertation. University of Illinois, Chicago. https://indigo.uic.edu/articles/Science_Education_as_a_Catalyst_for_Social_Change_Justice-Centered_Pedagogy_in_Secondary_Chemistry/10896803/1

National Research Council. (2012). *A Framework for K-12 Science Education: Practice, Cross Cutting Concepts, Core Ideas (Division of Behavioral and Social Sciences and Education)*. Washington, DC: National Research Council.

NGSS Lead States. (2013). *Next Generation Science Standards: For States, by States*. Washington, DC: The National Academies Press.

Paus, T. (2005). Mapping Brain Maturation and Cognitive Development during Adolescence. *TRENDS in Cognitive Sciences*, *9*, 60–68.

Paus, T. (2013). How Environment and Genes Shape the Adolescent Brain. *Hormones and Behavior*, *64*(2), 195–202. https://doi.org/10.1016/j.yhbeh.2013.04.004. Epub April 23, 2013. PMID: 23624012.

Vera-Estay, E., Dooley, J.J., & Beauchamp, M.H. (2015). Cognitive Underpinnings of Moral Reasoning in Adolescence: The Contribution of Executive Functions. *Journal of Moral Education*, *44*(1), 17–33. https://doi.org/10.1080/03057240.2014.986077

# Part II
# Integration of Anti-Racism Into High School Biology Curricula
## Lessons

### Introductory Thoughts

Here follow a suite of methods, exercises, and activities, that are designed to facilitate the introduction of the topic of social justice into biology classrooms. In short, these chapters (2–4) can be thought of as practical guides for introducing the approach we advocate in this book. The prescriptions presented in them are designed to be utilized within the context of the biology evolution/natural selection sections of biology courses, since it is here that a deeper understanding of evolution is explored, and it is with these concepts that racial justice is most obviously and easily connected to evolutionary. Four pedagogical alignments of instruction are presented for each lesson: Next Generation Science Standards (NGSS); College Board (AP Biology); Vision and Change; and the 5E model. The lessons described in this book can be addressed any time during the academic year in a variety of biology classes: honors, college prep, and AP, and do not need to be restricted to the evolutionary biology parts of these classes. For example, the historical oppression of women can be highlighted when exploring the work of Lynn Margulis (1938–2011), the originator of the then-radical endosymbiotic theory for the origin of eukaryotic cells; and when reviewing gene expression and regulation, students could learn about Rosalind Franklin (1920–1958), whose contribution to the discovery of the structure of DNA arguably should have earned her a spot in the shared Nobel Prize for Physiology or Medicine in 1962. In short, these ideas need not be presented as a single unit, but could (and perhaps should) be revisited again and again as appropriate opportunities open up in the curriculum. Any teacher dedicated to these ideas will find much fertile ground here.

DOI: 10.4324/9781003409144-3

## Reviewing Racially Based Social Injustices

The prescriptions in the following three chapters require students to familiarize themselves with at least some of the fraught US history of racially based social injustices. An extended treatment of what is meant this is presented in Chapter 5 of this book. Racial injustices can arise from laws that intentionally (explicitly) or unintentionally (implicitly) create the conditions which oppress certain groups (see Chapters 8–10). Currently, in the United States, race-related social injustices include inequities in voting access, treatment of the judicial system, policing laws, environmental issues (i.e., clean water and air, industrial waste disposal), health care laws (i.e., insurance mandates and coverage eligibility), education access (i.e., public school segregation, and labor access (i.e., worker's rights, occupational health and safety; see Chapter 9).

Our society embraces the idea of meritocracy, a system that rewards merit (ability + effort) with success. However, data show that success is not the result of only intelligence and skills; instead, family wealth and social position play a key role in getting ahead. This becomes problematic when wealth and status have been unevenly distributed based on repressive systems that favor some over others, irrespective of intelligence, skills or effort. Social injustice can be based on many factors, however, these lessons focus specifically on "race" and the ways in which the grouping of humans has played a role in elevating some people while subjugating others (see Chapter 10).

Similarly, when addressing the impact of global climate change with students, they ought to be exposed to the historical events that have resulted in disproportionate impacts on Black, Indigenous, and People of Color (BIPOC). Higher temperatures are, perhaps unexpectedly, linked to reductions in standardized test scores, an educational outcome with consequential results, particularly for Black and Latinx children (Park et al., 2020). The inequities in access to learning environments that are conducive to learning, such as air-conditioning during hot days, show up as limiting factors for BIPOC students. Likewise, redlining—the systematic refusal of loans or insurance in areas deemed to be a financial risk due to race (Chapter 9)—resulted in many Black neighborhoods today with more paved areas and fewer trees. In turn, those neighborhoods are now hotter than their white counterparts, leading to more cases of heat-related illnesses (Hoffman et al., 2016) and a paucity of medical support. The disparities for BIPOC may begin as early as *in utero*, since pregnant women exposed to high temperatures or air pollution are more likely to have children who are premature, underweight, or stillborn, and BIPOC mothers and babies are harmed at a much higher rate than the population at large (Bekkar et al., 2020).

Students will respond more readily to certain subjects. One way to have students explore data is to present data as "bell-ringers/Do now" topics.

These data can be displayed and students can be asked about what they notice and/or what questions they have. Here, we offer some suggestions for "high-impact" data; ideas with which experience suggests that students' interests will resonate. These can be grouped into separate categories:

## Historical Data

- Following Christopher Columbus' arrival to the Americas in 1492, violence and disease killed 90% of the indigenous population—nearly 55 million people.
- From the 16th to the 19th century during the transatlantic slave trade, 10 to 12 million African men, women, and children were kidnapped and enslaved and brought across the Atlantic Ocean to the Americas.
- The Holocaust was the genocide of the European Jews between 1941 and 1945. During World War II Nazi Germany and its collaborators systematically murdered some six million Jews, around two-thirds of Europe's Jewish population.
- Colonialism shaped immense economic inequality and the plundering of wealth and human capital from the colonized places helped much of the West get rich. Currently, the West drains commodities from the former colonies worth $2.2 trillion per year. For perspective, that amount of money would be enough to end extreme poverty, globally, fifteen times over.
- Even today, western-style amenities (e.g., 5-star hotels and restaurants in non-western locales) can be seen as a kind of modern colonialism in which tourist dollars, so necessary for the local economies, ultimately impinge upon the lifestyles, culture, and environments of indigenous inhabitants.

Cited Works for bulleted data above

Hickel, J., Sullivan, D., & Zoomkawala, H. (2021). Plunder in the Post-Colonial Era: Quantifying Drain from the Global South Through Unequal Exchange, 1960–2018. *New Political Economy, 26*(6), 1030–1047. https://doi.org/10.1080/13563467.2021.1899153

Hoffman, K.M., Trawalter, S., Axt, J.R., & Norman Oliver, M. (2016). Racial Bias in Pain Assessment and Treatment Recommendations, and False Beliefs About Biological Differences Between Blacks and Whites. *Proceedings of the National Academy of Sciences, 113*(16), 4296–4301.

Koch, A., Brierley, C., Maslin, M., & Lewis, S. (2019, February 1). *European Colonisation of the Americas Might Have Caused Global Cooling, According to New Research*. World Economic Forum. Retrieved December 21, 2022, from www.weforum.org/agenda/2019/02/european-colonisation-of-the-americas-caused-global-cooling/

Meredith, M. (2014). *The Fortunes of Africa*, p. 191. New York: Public Affairs.

United States Holocaust Memorial Museum. Introduction to the Holocaust. *Holocaust Encyclopedia*. Retrieved December 21, 2022, from https://encyclopedia.ushmm.org/content/en/article/introduction-to-the-holocaust

Some students respond best to data that show modern social injustice based on race in many different areas of life including housing, healthcare access, treatment by the judicial system, educational opportunities and generational wealth (see Chapter 9).

## Modern Data

- Housing: in 2022, 45.3% of Black households own their own home, compared with more than 74.6% of white households.
- Healthcare access: on average BIPOC (Black, Indigenous, and People Of Color), receive lower quality health care than white people and suffer higher mortality rates by almost all measures.
- Treatment by the judicial system: black youth were 269% more likely to be arrested for violating curfew laws than white youth.
- Education opportunities: two-thirds of BIPOC students attend schools that are predominantly BIPOC, most of them located in central cities and funded well below those in neighboring suburban districts.
- Wealth: in 2019 the net worth of a typical white family ($188,200), is nearly ten times greater than that of a Black family ($24,100).
- Limited representation in political power: White men in this country hold 62% of all elected offices despite being just 30% of the population.

Cited Works for bulleted data above

Bhutta, N., Chang, A.C., Dettling, L.J., & Hsu, J.W. (2020, September 28). *Disparities in Wealth by Race and Ethnicity in the 2019 Survey of Consumer Finances, FEDS Notes*. Washington: Board of Governors of the Federal Reserve System. https://doi.org/10.17016/2380-7172.2797

Bridges, K. (2018). *Implicit Bias and Racial Disparities in Health Care*. Retrieved November 2, 2022, from www.americanbar.org/groups/crsj/publications/human_rights_magazine_home/the-state-of-healthcare-in-the-united-states/racial-disparities-in-health-care/

Darling-Hammond, L. (2014). Closing the Achievement Gap: A Systemic View. In J. Clark (Ed.), *Closing the Achievement Gap from an International Perspective*. Dordrecht: Springer. https://doi.org/10.1007/978-94-007-4357-1_2
Henderson, T. (2022, October 13). *Black Families Fall Further Behind on homeownership*. The Pew Charitable Trusts. Retrieved December 21, 2022, from www.pewtrusts.org/en/research-and-analysis/blogs/stateline/2022/10/13/black-families-fall-further-behind-on-homeownership#:~:text=In%202022%2C%2074.6%25%20of%20White,%2C%20a%2027%2Dpoint%20gap
Research and Analysis of the 2020 Primary Elections. Reflective Democracy Campaign. (2022, April 23). Retrieved December 21, 2022, from https://wholeads.us/research/system-failure-2020-primary-elections/
Rovner, J. (2014). *Disproportionate Minority Contact in the Juvenile Justice System*. The Sentencing Project. www.jstor.org/stable/resrep27325

Finally, teachers may find that examples of race-based disparity can be effectively highlighted by exploring recent events which impact whites differently from non-whites.

## Current Data

- Water crisis in Flint, Michigan. As a result of systemic racism, inadequate treatment and testing of the water resulted in a series of major water quality and health issues for Flint residents (majority Black); chronically ignored, overlooked, misled, and ultimately discounted by government officials.
- COVID-19. There are no biological factors that make BIPOC more likely to be affected by COVID-19. Yet, BIPOC are disproportionately affected by COVID-19 at all stages of the disease from exposure to healthcare treatment, to hospitalizations, to job and wealth losses, to death.
- Access to healthcare. The National Academy of Medicine (NAM) reported that Black people are disproportionately sicker and have shorter life spans than their white complements. NAM found that "racial and ethnic minorities receive lower-quality health care than white people—even when insurance status, income, age, and severity of conditions are comparable." BIPOC are less likely than white persons to be given appropriate cardiac care, to receive kidney dialysis or transplants, and to receive the best treatments for stroke, cancer, or AIDS.

Cited Works for bulleted data above

Bekkar, B., Pacheco, S., Basu, R., & DeNicola, N. (2020). Association of Air Pollution and Heat Exposure with Preterm Birth, Low Birth Weight, and Stillbirth in the US: A Systematic Review. *JAMA Network Open, 3*(6), e208243. https://doi.org/10.1001/jamanetworkopen.2020.8243

Bridges, K. (2018). *Implicit Bias and Racial Disparities in Health Care*. Retrieved November 2, 2022, from www.americanbar.org/groups/crsj/publications/human_rights_magazine_home/the-state-of-healthcare-in-the-united-states/racial-disparities-in-health-care/

Centers for Disease Control and Prevention. (2020). *Health Equity Considerations and Racial and Ethnic Minority Groups*. Retrieved November 4, 2022, from www.cdc.gov/coronavirus/2019-ncov/community/health-equity/race-ethnicity.html

Michigan Civil Rights Commission. (2017). *The Flint Water Crisis: Systemic Racism Through the Lens of Flint*. Retrieved November 4, 2022, from www.michigan.gov/documents/mdcr/VFlintCrisisRep-F-Edited3-13-17_554317_7.pdf

Park, R.J., Goodman, J., & Behrer, A.P. (2020). Learning is Inhibited by Heat Exposure, Both Internationally and within the United States. *Nature Human Behaviour*. https://doi.org/10.1038/s41562-020-00959-9

# 2

# Lessons on Race and Biological Grouping

<table>
<tr><th colspan="2">Next Generation Science Standards</th></tr>
<tr><td>Science & Engineering Practices:<br>Constructing Explanations & Designing Solutions<br>Engaging in Argument from Evidence</td><td>Cross Cutting Concepts:<br>Structure and Function<br>Cause and Effect</td></tr>
<tr><td colspan="2">Supported NGSS Standards:<br>HS-LS3-2 Heredity–Inheritable Genetic Variation: Make and defend a claim based on evidence that inheritable genetic variations may result from: (1) new genetic combinations through meiosis, (2) viable errors occurring during replication, and/or (3) mutations caused by environmental factors. (Cause and Effect)<br><br>HS-LS4-2 Biological Evolution–Four Factors of Natural Selection: Construct an explanation based on evidence that the process of evolution primarily results from four factors: (1) the potential for a species to increase in number, (2) the heritable genetic variation of individuals in a species due to mutation and sexual reproduction, (3) competition for limited resources, and (4) the proliferation of those organisms that are better able to survive and reproduce in the environment. (Cause and Effect)<br><br>HS-LS4-4 Biological Evolution–Natural Selection Leads to Adaptation. Construct an explanation based on evidence for how natural selection leads to the adaptation of populations. (Cause and Effect)</td></tr>
</table>

DOI: 10.4324/9781003409144-4

<table>
<tr><td colspan="2">Disciplinary Core Ideas:<br>LS3.B: Variation of Traits; LS4.B: Natural Selection; LS4.C: Adaptation</td></tr>
<tr><td colspan="2" align="center">AP Biology</td></tr>
<tr><td colspan="2">Essential knowledge:<br>EVO-1.E.1: Natural selection acts on phenotypic variations in populations.<br>EVO-1.E.2: Environments change and apply selective pressures to populations.<br>EVO-1.E.3: Some phenotypic variations significantly increase or decrease fitness of the organism in particular environments.<br>EVO-1.I.1: Reduction of genetic variation within a given population can increase the differences between populations of the same species.<br>EVO-1.N.2: A comparison of DNA nucleotide sequences and/or protein amino acid sequences provides evidence for evolution and common ancestry.<br>SYI-3.D.1: The level of variation in a population affects population dynamics<br>IST-1.J.1: Patterns of inheritance of many traits do not follow ratios predicted by Mendel's laws and can be identified by quantitative analysis, where observed phenotypic ratios statistically differ from the predicted ratios.<br>IST-1.J.3: Many traits are the product of multiple genes and/or physiological processes acting in combination; these traits therefore do not segregate in Mendelian patterns.<br>SYI-3.B.1: Environmental factors influence gene expression and can lead to phenotypic plasticity. Phenotypic plasticity occurs when individuals with the same genotype exhibit different phenotypes in different environments.</td></tr>
<tr><td colspan="2" align="center">Vision and Change</td></tr>
<tr><td>Core Competencies:<br>1) the ability to apply the process of science; 4) the ability to tap into the interdisciplinary nature of science; 5) the ability to communicate and collaborate with other disciplines; and 6) the ability to understand relationships between science and society.</td><td>Core Concepts:<br>1) evolution (the diversity of life-forms that have evolved over time through mutations, selection and genetic change)</td></tr>
</table>

As an introduction to the activities in these lessons, please check the overview videos of the authors, and other experts, as these videos provide an overview of the content and the philosophy behind the material. The video for this chapter can be found here: https://youtu.be/JZXIy9iD6Og

## Activity 1: Defining Race and Racism (See Chapter 6—The Roots of Oppression: Slavery, Colonization, and Unfettered Free Markets)

### Overview

Students will begin defining race and analyzing data related to racism to understand its importance.

Suggested Timing: 45–60 minutes
Materials: Student notebook, a way for students to view videos.

Prerequisite Knowledge: Students should be familiar with basic genetic heredity concepts, including information molecules (e.g., DNA), as well as the definition of biological evolution.

### Driving Questions

What is race? Why are there differences in skin coloration in humans? How do genetic factors contribute to skin color differences? What are the impacts of creating a racial hierarchy? Why are we talking about race in our biology class?

### 5E Model of Activity 1

| Sequence of events | Notes |
|---|---|
| Engage<br>"What do I already know about this?" | Students think about what race is (and what it is not). |
| Explore<br>"Discusses problems with others" | Students share their definitions and what they know about the key terms. |
| Explain<br>"Compares current thinking with former thinking" | Students explain how social groupings of people based upon specious biological principles can have a negative impact on life outcomes. |

| Sequence of events | Notes |
|---|---|
| Elaborate<br>"Uses previous information to ask questions, propose solutions, make decisions, design experiments, or complete a challenge" | Students create a list of ways to address the question, "If scientists were interested in the existence of racism, what kind of data could they explore?" |
| Evaluate<br>"Assesses progress by comparing current understanding with prior knowledge" | Students generate research questions based on their understanding of what racism is and the interrelationship among race and life outcomes and conduct research to see if any of these topics have been explored scientifically. |

## Engage

Begin the activity by asking students the question "What is race?" Record possible answers on the board.

## Explore

Play the following video segment from the Howard Hughes Medical Institute: www.biointeractive.org/classroom-resources/biology-skin-color (0:00 to 9:00 min. and 16:06 to 17:58).

In a large discussion group, have students share their definitions and what they know about the following terms. Be sure to provide formal definitions for students (Race, Racism, White supremacy).

Have the entire class revisit the concept of race and modify their definition. Ask students to respond to questions about content from the video:

What is race and how is it related to skin pigmentation patterns around the world?
What are the selective pressures that gave rise to different pigmentation patterns around the globe?
What are the results of exclusively grouping people together by skin pigmentation patterns?
What does modern science say about the human species and the idea of race?

## Explain

Now show the video "The Science of Racism," from AsapSCIENCE https://youtu.be/rBWImda1Zig (6 min.), which introduces the relationship between racism and negative life outcomes.

Now show the video "Legacies of Racial Slavery (3 min.) www.choices.edu/video/legacies-of-racial-slavery/

Students should start to see how social groupings of people based upon specious biological principles can have a negative impact on life outcomes.

Then, ask the students to answer the questions below.

How can we describe the relationship between race and racism?

How are the relationships you described in the previous question relevant to today's society? What inference can you make about grouping humans based on skin pigmentation patterns and social constructs that oppress certain people?

*Teacher Tip:*
*As you guide your students to think about racial grouping and the science used to validate them, the ideas described in this quote may be useful,*

> *"Racial categories historically developed and endorsed by scientists led to a hierarchy of groups seen as superior or inferior. Although unsupported by biological evidence, these categories have had devastating effects on non-white communities throughout history. The myth that racial groups were fundamentally different was used to justify colonialism, slavery, genocide, and eugenics, and it still governs policies today. The intersectionality of racism and modern society has left a legacy of racial disparities in socioeconomic status, education, and health."*
>
> Odekunle, E.A. (2020).

Dismantling Systemic Racism in Science. *Science Magazine, 369*(6505), 780–781. https://science-sciencemag-org.uri.idm.oclc.org/content/sci/369/6505/780.3.full.pdf

*An additional video that describes discrimination as a philosophical dilemma can be shown: Crash Course Philosophy #41 (9 min.) https://youtu.be/RsldtV4jWA0*

## Elaborate

Have students list data that could be explored in order to have a sense of racism (See Chapter 9—Oppression in the United States). This can be done by listing student ideas on the board to the question, "If scientists were interested in the existence of racism, what kind of data could they explore?"

## Evaluate (Check for Understanding)

Next, have students generate research questions based on their understanding of what racism is and the interrelationship between race and life outcomes. Finally, have students search the internet to see if their questions have been addressed by scientific research.

This one-point rubric can be used to assess students' questions and learning.

| Needs work | Criteria | Amazing parts |
|---|---|---|
| | **Relevance:** The questions are directly related to the content of the reading & demonstrate understanding of the main points. | |
| | **Depth:** The questions go beyond surface-level understanding & probe more deeply into the content of the reading. | |
| | **Breadth:** The questions cover a range of topics & perspectives related to the reading, rather than focusing on just one aspect. | |
| | **Creativity:** The questions show originality & insight, and encourage critical thinking about the material. | |
| | **Clarity:** The questions are clearly worded & easy to understand. | |
| | **Quantity:** The student generates a sufficient number of questions to demonstrate a thorough understanding of the reading. | |
| | **Quality:** The student generates high-quality questions that are thought-provoking & engage with the material in a meaningful way. | |

## Activity 2: Biological Grouping

### Overview

In this activity, students will explore how humans can actually be grouped based on biological data.

Suggested Timing: 45–60 minutes
Materials: Student notebook, printed quotes for each student (from Fuentes (2021).

Prerequisite Knowledge: Students should be familiar with basic grouping schemas (ways that people have grouped living organisms).

### Driving Questions

How are living organisms grouped? Why are humans grouped in specific ways? How are humans actually grouped according to biological data?

### 5E Model of Activity 2

| Sequence of events | Notes |
|---|---|
| Engage<br>"Demonstrates engagement by expressing ideas, sharing observations, and creating initial models" | Students read the two quotes from Fuentes (2021) on human racial grouping and create an entry that explains "human grouping" for their biology textbook. |
| Explore<br>"Compares ideas with those of others" | Students share their textbook entries. |
| Explain<br>"Adjusts ideas, models, and explanations as new evidence or reasoning is presented" | Students read the article "Ten Facts about Human Variation." |
| Elaborate<br>"Uses previous information to ask questions, propose solutions, make decisions, design experiments, or complete a challenge" | Students participate in a "jigsaw" activity, based on the essay by Marks (2010). |
| Evaluate<br>"Assesses progress by comparing current understanding with prior knowledge" | Students participate in a gallery walk about the topics discussed in the lesson. |

### Quote 1

Substantive evidence from paleoanthropology, human biology, biological anthropology, and genomics demonstrate clearly and unequivocally that all contemporary humans belong to one species and one subspecies: *Homo sapiens sapiens*. Across the evolution of the genus *Homo* over the past ~2.5 million years, there were multiple lineages, possibly different species, and different subspecies (e.g., Neanderthals, Denisovans, *Homo erectus*, *Homo naledi*, etc.). However, for at least the past ~35,000 years, there has been only one kind of human on this planet, us—*Homo sapiens sapiens* (p. 156).

### Quote 2

"Race does not provide an accurate representation of human biological variation. It was never accurate in the past, and it remains inaccurate when referencing contemporary human populations. Humans are not divided biologically into distinct continental types or racial genetic clusters. Instead, the Western concept of race must be understood as a classification system that emerged from, and in support of, European colonialism, oppression, and discrimination. It thus does not have its roots in biological reality but in policies of discrimination."

American Association of Biological Anthropologists (AABA) Statement on Race and Racism (p. 156).

## Engage

For homework, assign students to read the two quotes from Fuentes (2021) on human racial grouping and ask them to create an entry that explains "human grouping" for their biology textbook. Students should compare "old" (Linnean) methods and modern methods for grouping humans and explain how different ways of grouping humans affect our society. This entry should be between 1 and 2 pages, and students should be encouraged to include meaningful text illustrations, images, and other data (all appropriately cited).

## Explore

In class, the following day, discuss the entries created by students and check for understanding by calling on different students to explain the different reasons they chose specific text and images.

Then, in groups, have students brainstorm examples of the negative impact of racism in their community. Subsequently, as a class watch this video (3 min.): Humans and Race: Exploring the Intersections of Science and Racism (https://youtu.be/TRzusl6hVQs). Additional related materials can be found here: https://sciencereligiondialogue.org/resources/humans-and-race/

## Explain

Now students should read the article "Ten Facts about Human Variation," by J. Marks (https://webpages.uncc.edu/~jmarks/pubs/tenfacts.pdf). Marks, J. (2010). Ten Facts about Human Variation. In M.P. Muehlenbein (Ed.), *Human Evolutionary Biology* (pp. 265–276). Cambridge: Cambridge University Press.

In order to speed up the process, as a class read the introduction collectively, then assign each student (or pair of students, depending on the size of the class) a specific "fact" to read and report back to the whole class.

## Elaborate

Students participate in a "jigsaw" activity, where each student in a group reads a different text or passage, taking on the role of "expert" on what was read. Students share the information from that reading with students from other groups and then return to their original group to share their new knowledge.

The ten facts that Marks lists are:

1. Human groups distinguish themselves principally culturally.
2. Human biological variation is continuous, not discrete.
3. Clustering populations is arbitrary.
4. Populations are biologically real, not races.
5. Populations also have a constructed component.
6. There is much more variation within groups. (polymorphism) than between groups (polytypy).
7. People are similar to those nearby and different from those far away.
8. Racial classification is historical and political and does not reflect natural biological patterns.
9. Humans have little genetic variation.
10. Racial issues are social-political-economic, not biological. (Marks, 2010, pp. 265–276)

*Teacher Tip:*
*An additional source that explains the need for this type of lesson is found here: https://scicomm.plos.org/2018/11/29/it-is-unethical-to-teach-evolution-without-confronting-racism-and-sexism/*

*If time allows, after students read the second article on the ten facts of human variation, show the entire class the TED talk, "John Biewen: The lie that invented racism" https://youtu.be/oIZDtqWX6Fk (18 min.) and be prepared to contribute to a class discussion that addresses the two questions.*

*Discussion Questions:*

1. *John Biewen argues that racism is still with us today. Explain why he feels this way. What data does he cite?*
2. *Why does Biewen say that "whiteness" is a story? Who invented the idea of race and for what purpose?*

This discussion will provide the contextual framework that students will use to evaluate solutions for racism. You can assign this video as homework and then use the following class period as a flipped class to discuss and clear up misconceptions. The problems and solutions are complex and students must be able to extend their knowledge.

### Evaluate (Check for Understanding)

After reading the above article, conduct a Gallery Walk (The Teacher Toolkit) activity (described below) based on the "discussion points" from the article.

A. Write each of these questions on a piece of chart paper or on a whiteboard. Hang or place the questions or prompts in various places around the classroom to create six stations.
   1. **What are the incompatibilities among the three/four concepts of race discussed in this essay?**
      *Potential answer: according to the author: essentialist, population, genetic residual, and the biocultural construction of race. Students can focus on any of the three listed concepts of race, 1. The human species could be naturally partitioned into a reasonably small number of reasonably discrete kinds of people, 2. Race is a group of people, a population, rather than an inner quality or spirit (rather than race residing within a person, a person is part of a race), and 3. For population geneticists, race is people from a geographically localized gene pool or population (based on the small amount of difference detectable among the most geographically separated peoples). Students can then point out all of the issues with these concepts (e.g., human groups distinguish themselves principally culturally (not biologically); human biological variation is continuous, not discrete; clustering*

*populations is arbitrary; racial classification is historical and political, and does not reflect natural biological patterns; humans have little genetic variation; racial issues are social-political economic, not biological, etc.).*

2. **Why can't we separate facts of nature from culture?**
*Potential answer: Simply put, because they aren't additive. As stated in the article, "the vast bulk of human behavioral and mental diversity is culturally constituted." Interactions between the "natural" and the "cultural" are complex and problematic to separate since historical events often have biological consequences, which in turn engender differences in responses.*
3. **Are "Hispanics" a race?**
*Potential answer: No. Hispanic (a term used in the 1980 US Census) is a term that describes a large group of people who speak Spanish which is just one ancestors' language (according to the US Census, "a person of Cuban, Mexican, Puerto Rican, South or Central American, or other Spanish culture or origin regardless of race."). Similar to the term Latino (or Latinx), these categories are descriptive of cultural backgrounds not genetic grouping (as Latinos have a broad range of genetic histories).*
4. **Old anthropology books used to show maps of the races of the world, with, for example, no presence of Europeans, Asians, or Africans in America. What are the merits of, and problems with, that?**
*Potential answer: There are few merits to the idea that humans are static and permanent residents around the world. Humans, like other migratory species, have been migrating around the planet for at least 65,000 years since the first humans left Africa. Moreover, currently, there are human interactions and exchanges of peoples, ideas, and genes.*
5. **What are the major patterns of human genetic variation and the major patterns of human cognitive variation, and how do they relate to one another? What implications can be drawn from that?** (Marks, 2010, p. 274)
*Potential answer: Human cognitive display is regulated by culture (and the historical facts that have elevated some people over others). As pointed out in the article, "Lewontin numbers show that patterns of human genetic diversity simply do not map well onto the patterns of human behavioral or cognitive diversity. The latter kinds*

*of differences tend to be localized at the borders of human groups, as noted above, and are of the sort we call cultural." Biological (genetic) variations cannot explain the historical and socio-political trends of our society.*

Cited Works for articles used in this lesson

Fuentes, A. (2021). "On the Races of Man": Race, Racism, Science, and Hope. In J. Desilva & J. Browne (Eds.), *A Most Interesting Problem: What Darwin's Descent of Man Got Right and Wrong About Human Evolution*, pp. 144–161. Princeton, NJ: Princeton University Press.

Marks, J. (2010). Ten Facts About Human Variation. In M.P. Muehlenbein (Ed.), *Human Evolutionary Biology*, pp. 265–276. Cambridge: Cambridge University Press.

6. **Print this image and write below it, "What is wrong with this image?"**

**Figure 2.1** Illustration of Supposed Human Evolution

*Potential answer: This image incorrectly shows a direct, linear pathway for human development. Similarly, there are racist undertones that show up as the more developed (evolved) human is usually portrayed as the lightest hominid on the diagram. Finally, the ancestor of* Homo sapiens *was not a chimpanzee; chimpanzees and humans share a common ancestor, from which they diverged some ~6 million years ago.*

**7. Print this image about human "races," and write the question, "what does this image show?"**

**Figure 2.2** Types of Human Race

*This illustration is drawn from a textbook, Geography for beginners, published in 1864 and intended for use in the schools of the Confederate southern United States.*

David Pearson, Ethnic Inequalities—Race, Ethnicity and Inequality. *Te Ara—The Encyclopedia of New Zealand*. Retrieved

December 7, 2022. www.TeAra.govt.nz/en/artwork/28375/types-of-the-human-race
*Potential answer: This type of image conveys similar ideas to Linnaeus where the white man is displayed as the well-educated and most developed human. The position within the diagram (the center), the clothing, and even the angle of the face, all point to the idea that the white man is superior to the others.*

B. Group students into teams of three to five students, and each group should start at a different station.
C. At the first station, groups will read what is posted and one recorder should write the group's responses, thoughts, and comments on the chart paper or whiteboard. For individual student accountability, you may also have the students record their own responses on a worksheet, or put their initials below what they wrote. Having different colored markers for each student is also an option.
D. After three to five minutes, have the groups rotate to the next station. Students read and discuss the previous group's response and add content of their own. Repeat until all groups have visited each station. To involve all group members, you can have groups switch recorders at each station.
E. As the teacher, it is important to monitor the stations while the students participate. You may also need to clarify or provide hints if students don't understand or misinterpret what is posted at their station.
F. Have students go back to their first station to read all that was added to their first response. Bring the class back together to discuss what was learned and make final conclusions about what they saw and discussed.

As a final assessment for this activity, students can reflect by answering these questions individually:

A. What did you like about Gallery Walk?
B. What did you dislike about Gallery Walk?
C. How did the gallery walk compare to the usual class discussion techniques? Think about the sharing of good ideas, inspiring others; clearly communicating, etc.
D. How would you rate your participation? How about the participation of classmates?
E. In what ways did your classmates' ideas help you to refine your thinking?

Alternatively, students can read this list of ten facts about human variation from *Racism, Not Race: Answers To Frequently Asked Questions*. By Joseph L. Graves Jr. and Alan H. Goodman. (2021). Columbia University Press. (ISBN-10: 0231200668)

1. Human biological variation is real. It is patterned, important, and a thing of beauty to be celebrated.
2. Race neither describes nor explains human biological variation.
3. Humans do not have biological races.
4. Human biological races are a relatively recent idea that was reified—made real legally and scientifically—to justify racism.
5. Racial classifications developed historically as a politically important means to categorize and divide individuals.
6. Many individuals still believe in the myth of race as being "obviously" biological, in the blood and genes. Race is a powerful illusion. That myth provides ideological justification for systemic racial inequalities.
7. Racism is an ideology that is built on a myth and widely shared, with institutional and structural manifestations.
8. The proof of the impact of racism is found in the data on inequalities in almost all aspects of life, including education, employment, health, and wealth.
9. Race will become less salient when racial ideology is overcome and races reach equality in measures of life such as health and wealth.
10. We cannot have a civil and just society without racial equality. Racial equality will be good for everyone. (p. 237–238)

In small groups, students can create a mini poster/illustration about one of these facts and share their products with the whole class. The rubric below can be used to assess student learning.

| **Areas that need work** | **Criteria** | **Things that were amazing** |
|---|---|---|
| | The mini poster/illustration effectively communicates a clear and concise message, using appropriate visuals and layout. | |

## Activity 3: What Does Science Say About Human Grouping? Working Toward Solutions

### Overview

In this lesson, students will explore what science and evolutionary theory actually say about race, then begin to solidify an action plan. This lesson has students move beyond the problems created by racism to researching solutions that can reduce and remediate these problems. Students will read a couple of articles, watch a couple of short videos, and discuss how learning about biological grouping could be used to raise awareness of racism and better understand how scientific knowledge could be used to reduce race-based oppression. By the end of the lesson, students will identify local, national, and global efforts to combat racism and explore ways in which local organizations are working on improving the life outcomes of people who suffered due to racism.

*Teacher Tip:*
*Current scientific understanding is clear about the unity of the human species. Students should consider how the data which show the general homogeneity of the human species have been collected, vetted, interpreted, and validated.*

Suggested Timing: 45–90 minutes.
Materials: Student notebook, internet access, means of watching the videos.

### Driving Questions

What do science and evolutionary theory actually say about race? How can the lack of evidence for racial divisions help us to create a more just society? What are some potential solutions to racism?

### 5E Model of Activity 3

| Sequence of events | Notes |
|---|---|
| Engage<br>"Shows interest in the topic through curiosity and expression of wonderings" | Students watch the "Race & Biology video by Dr. Graves Jr. (~9 min). |
| Explore<br>"Compares ideas with those of others" | Students read the "How Science and Genetics are Reshaping the Race Debate of the 21st Century" article |

| Sequence of events | Notes |
|---|---|
| Explain<br>"Explains using evidence from investigations" | Students create an infographic. |
| Elaborate<br>"Makes conceptual connections between new and previous experiences" | Students learn about the concepts researched in the "Are Rats Born Racist?" study. |
| Evaluate<br>"Assesses progress by comparing current understanding with prior knowledge" | Students write for a short, specific amount of time about a designated topic related to a prompt. |

## Engage

Have students watch the "Race & Biology video by Dr. Graves Jr.: https://vimeo.com/240722713 (8:45 min.). Ask students to reflect on what they have learned by sharing their ideas in a smaller group, and eventually reporting to the whole group (by a representative of the group).

Students should mention some of the following main ideas in their reflections:

- The pernicious effects of racial thinking have led to the abuse and oppression of certain groups deemed inferior, such as colonialism, genocides, and enslavement.
- Modern scientists have found that differences between human groups are minimal and not biologically significant.
- Biological determinism is a fairly new set of ideas that started around the time of European colonization.
- Racial medicine erroneously credits biological differences between groups of people for the health outcome disparities which impact BIPOC.
- There are biological races among some animals, but not in humans. This is likely due to the fact that we are a fairly young species (around 200,000 years old) and there is significant gene flow as a result of travel and commerce between human groups in different regions, and thus no biological reason for biological races to develop.

## Explore/Explain

Next, have students read the "How Science and Genetics are Reshaping the Race Debate of the 21st Century" article: http://sitn.hms.harvard.edu/flash/2017/science-genetics-reshaping-race-debate-21st-century/

The article has 5 short sections:

- Race in the new era of human genetics research
- New findings in genetics tear down old ideas about race
- Does "race" still mean something?
- Science and genetics: Instruments of modern racism
- Fighting racism with understanding

Since this article has five clearly delineated parts, it provides an opportunity for a shared reading activity, where small groups of students will read their assigned part, come up with a consensus about what it says and report to the whole class. These presentations can be done by making an "infographic" which explains the main points of their reading. Sample infographics can be searched on the internet, including those on the CDC website (www.cdc.gov/globalhealth/infographics/default.html).

## Elaborate

Finally, have students read the article "Are Rats Born Racist?" which shows that social contact matters more than genetics in determining who gets help and who doesn't: https://greatergood.berkeley.edu/article/item/are_rats or watch this short video: https://youtu.be/TNKbDTcDsqw or the original scientific article: https://elifesciences.org/articles/01385.

*Teacher Tip:*
*Depending on experience, some students can receive and utilize the data from primary sources such as scientific journal articles, such as: www.uchicagomedicine.org/forefront/news/social-experience-drives-empathetic-pro-social-behavior-in-rats*

## Evaluate (Check for Understanding)

Once students complete the reading, they will write for a short, specific amount of time about a designated topic related to a prompt. After they have finished reading, provide students with 20–25 minutes to complete a short writing exercise using the following questions as guidance.

- What do I already know about human differences and races?
- What is something new I learned about social behaviors in rats?
- In which ways is this information about rats changing their behaviors based on experiences relevant to humans?

- Does this information make sense to me?
- What were the most important points in this reading?
- How does it fit in with what I already know about racism in humans?
- Do I still have questions?

In a large group discussion, ask students to share any potential beneficial effects of addressing racism in their communities and the world. Be sure that students cite evidence from their readings/research.

As an assessment this single-point rubric can be used to analyze the individual writing:

| Areas that need work | Criteria | Things that were amazing |
|---|---|---|
| | **Overall Thinking/Writing:** The writing is clear, creative, and consistently supported by the evidence. | |
| | **Composition:** Focused on deep reflection and learning and included supporting details to explain logic, | |
| | **Written Expression:** Used appropriate, topic vocabulary. Used a variety of sentence structures and appropriate voicing. | |

Lastly, students can create a "textbook page" about "race." This page should include both the insidious history of racist ways of grouping humans, as well as the current scientific ways in which humans are actually grouped.

# 3

# Lessons On the History of Race-Based Oppression

| Next Generation Science Standards | |
|---|---|
| **Science & Engineering Practices:**<br>Asking Questions<br>Obtain, Evaluate, Communicate Information<br>Analyzing & Interpreting Data | **Cross Cutting Concepts:**<br>Patterns<br>Stability & Change<br>Cause & Effect: Mechanism & Explanation |
| **Supported NGSS Standards:**<br>*HS-LS2–8: Social Interactions and Group Behavior:* Evaluate the evidence for the role of group behavior on individual and species' chances to survive and reproduce. (Cause and Effect)<br><br>*HS-LS4–6: Human Impact on Biodiversity Solution:* Create or revise a simulation to test a solution to mitigate adverse impacts of human activity on biodiversity. (Cause and Effect)<br><br>*HS-LS4–2 Biological Evolution*–Four Factors of Natural Selection: Construct an explanation based on evidence that the process of evolution primarily results from four factors: (1) the potential for a species to increase in number, (2) the heritable genetic variation of individuals in a species due to mutation and sexual reproduction, (3) competition for limited resources, and (4) the proliferation of those organisms that are better able to survive and reproduce in the environment. (Cause and Effect) | |

DOI: 10.4324/9781003409144-5

<table>
<tr><td colspan="2">HS-LS4–4 Biological Evolution–Natural Selection Leads to Adaptation. Construct an explanation based on evidence for how natural selection leads to adaptation of populations. (Cause and Effect)</td></tr>
<tr><td colspan="2">Disciplinary Core Ideas:<br>LS4.D: Biodiversity and Humans; ETS1.B: Developing Possible Solutions; LS4.C: Adaptation</td></tr>
<tr><td colspan="2" align="center">AP Biology</td></tr>
<tr><td colspan="2">Essential knowledge:<br>EVO-1.H.1: Evolution is also driven by random occurrences:<br>a. Mutation is a random process that contributes to evolution.<br>b. Genetic drift is a nonselective process occurring in small populations—i. Bottlenecks; ii. Founder effect.<br>c. Migration/gene flow can drive evolution.<br>EVO-1.E.1: Natural selection acts on phenotypic variations in populations.<br>EVO-1.E.2: Environments change and apply selective pressures to populations.<br>EVO-1.E.3: Some phenotypic variations significantly increase or decrease fitness of the organism in particular environments.<br>EVO-1.I.1: Reduction of genetic variation within a given population can increase the differences between populations of the same species.<br>SYI-3.B.1: Environmental factors influence gene expression and can lead to phenotypic plasticity. Phenotypic plasticity occurs when individuals with the same genotype exhibit different phenotypes in different environments.</td></tr>
<tr><td colspan="2" align="center">Vision and Change</td></tr>
<tr><td>Core Competencies:<br>1) the ability to apply the process of science; 4) the ability to tap into the interdisciplinary nature of science; 5) the ability to communicate and collaborate with other disciplines; and 6) the ability to understand relationships between science and society.</td><td>Core Concepts:<br>1) evolution (the diversity of life-forms that have evolved over time through mutations, selection and genetic change. 5) systems (the ways in which living things are interconnected and interact with one another).</td></tr>
</table>

As an introduction to the activities in these lessons, please check the overview videos of the authors, and other experts, as these videos provide an overview of the content and the philosophy behind the material. The video for this chapter can be found here: https://youtu.be/yCJdgFVOFiQ

## Activity 4: The Historical Relationship between Racism and Science, Through Storytelling

### Overview

Now that students have a better understanding of what race and racism are, they will begin investigating their historical past and effects. Students will explore historical figures to learn more about the local and global effects of racism and science.

Students will describe different historical figures that were involved in the oppressive application of science; explain how skin coloration is a biological adaptation for humans based on UV radiation exposure; and describe how skin coloration was immorally used as a way to determine inferiority and the oppressive maltreatment of some groups.

Suggested Timing: 90–120 minutes (3–4 class periods)
Materials: Student notebook, internet access for researching their historical figure.

Prerequisite Knowledge: For context, students will benefit from having a general understanding of the timeline of American history including the initial arrival of enslaved Africans in 1619, the emancipation of slavery in 1865, Jim Crow era and legal segregation 1870–1954, Brown vs. the Board of Ed. in 1954 and the end of legal segregation in 1965 (see Chapter 7—Oppression in the United States).

**Driving Questions:** How have racists used science to attempt to validate their ideologies? How has science been misused to oppress some people?

#### Key Definitions

Scientific Racism: the scientific justification of racial discrimination and the scientific explanations offered in putative support for racial discrimination (see Chapter 8—Science and American racism).

#### Additional Resources

The United Nations provides a summary of the philosophy of scientific racism: www.un.org/en/chronicle/article/ideology-racism-misusing-science-justify-racial-discrimination

*Teacher Tip:*
*Students will be analyzing historical information which may be upsetting. It may be helpful to remind students that the purpose of the readings is to learn from the past, rather than to point fingers and incriminate any single individual or group of people. Additionally, it may be useful to explicitly tell students what their presentations should include (and even model one for them).*

## 5E Model of Activity 4

| **Sequence of events** | **Notes** |
|---|---|
| Engage<br>"What do I already know about this?" "What can I find out about this?" | Students provide examples of how scientists have behaved in racist ways. |
| Explore<br>"Plans and conducts investigations in which they observe, describe, and record data" | Students will explore the figures, then create a way to tell their story to the whole class. |
| Explain<br>"Explains using evidence from investigations" | Students create a short presentation about their story (5 min.). |
| Elaborate<br>"Critiques the models, explanations, or arguments made by others using evidence and reasoning" | Students provide feedback to the work of other students focusing on their story and asking questions regarding their historical figure. |
| Evaluate<br>"Assesses progress by comparing current understanding with prior knowledge" | Students answer questions related to the topic by writing a reflective abstract/essay (500 words or less) where the questions are addressed. |

## Engage

Ask students if they know any examples of how scientists have behaved in racist ways. Have a whole class discussion.

## Explore

Exploring historical figures that highlight racism, either by suffering from racist oppression or by promoting racism. Teachers highlight the fact that many of these people lived and were very active well into the 20th century. Students will explore the figures, then create a way to tell their story to the whole class. Have students conduct research on one of the figures listed and pay special attention to the way that race/racism played a role in their story (see Chapter 10—Science and American racism):

1. Carolus Linnaeus (1707–1778)
2. Samuel Cartwright (1793–1863)
3. Samuel Morton (1799–1851)
4. Josiah Clark Nott (1804–1873)
5. Louis Agassiz (1807–1873)
6. James Marion Sims (1813–1883)
7. Thomas Parran (1892–1968) [Tuskegee, Guatemala experiments]
8. Cornelius P. Rhoads Packard (1898–1959) [Puerto Rico]
9. Henrietta Lacks (1920–1951)
10. Rosalind Franklin (1920–1958)/James Watson (1928—present)

    -------------------------------------------

    (the names listed below can be used if needed)

11. David Starr Jordan (1851–1931)
12. Ferdinand Hayden (1829–1887)
13. Ernst Haeckel (1834–1919)
14. Rufus Bernhard von KleinSmid (1875–1964)
15. Charles Benedict Davenport (1866–1944)
16. Henry Herbert Goddard (1866–1957)

## Explain

Once students have documented their research findings, have students create a short presentation about their story (5 min.).

## Elaborate

Students provide feedback on the work of other students focusing on their story and asking questions regarding their historical figure. Feedback should include questions such as, "what did the presenter do well?," "in what ways can the presenter improve for next time?," "how does the historical figure covered by the presenter compare to my own historical figure?"

## Evaluate (Check for Understanding)

Individually, students will write a reflective abstract/essay (500 words or less) where the questions are addressed. Along with presenting historical information about their assigned figure, students will answer questions:

- What are the key points of the historical figure?
- How does your story relate to science?
- What is morally questionable about your story?
- What is scientifically questionable about your story?
- What lasting outcomes resulted from the story?
- How does your story show social injustice?
- Who was responsible for the injustice and to whom was it done?
- How can society address the injustice in your story?

*Teacher Tip:*
*Student presentations can be done for the same class (as students will only know the historical figures assigned to them), or for different classes, younger students, or even community members. A quick summary of these historical figures will be included for your review.*

A single-point rubric can be used to assess student writing:

| Areas that need work | Criteria | Things that were amazing |
|---|---|---|
| | **Thesis Statement** The thesis statement is clear, addresses the prompt, and makes an interesting argument. | |
| | **Overall Flow** The essay flows and logic is easy to follow. Arguments are clear, and supported by evidence. | |

| Areas that need work | Criteria | Things that were amazing |
|---|---|---|
| | **Grammar And Formatting** Appropriate grammar and formatting were used in the essay, correctly citing sources that were used. | |

Additionally, this rubric can be used to assess the student presentations:

| Areas that need work | Criteria | Things that were amazing |
|---|---|---|
| | **Content/Organization** Presentation is well-organized, with a clear introduction, body, and conclusion. Answered all questions with explanations and elaboration. | |
| | **Delivery** Used direct eye contact (seldom looking at notes). Spoke with fluctuation in volume and inflection to maintain audience interest and emphasized key points. Significantly increased audience understanding and knowledge of the topic. | |

## Activity 5: Migration Lessons (for Full Lesson and Rationale See https://knowlesteachers.org/kaleidoscope/unpacking-human-migration)

### Overview

Utilizing the tools and practices of science (asking questions, planning and carrying out investigations, analyzing and interpreting data, engaging in argument from evidence, obtaining, evaluating, and communicating information), students will explore the political rhetoric of human migration and its biological validity.

Suggested Timing: 90–120 minutes (3–4 class periods)
Materials: Student notebook, internet access for research

Prerequisite Knowledge: It is helpful for students to have some familiarity with biological migration and the politized discussions around human migration.

**Driving Questions:** Why do humans migrate? What are the effects of migration?

### 5E Model of Activity 5

| **Sequence of events** | **Notes** |
|---|---|
| Engage<br>"Shows interest in the topic through curiosity and expression of wonderings" | Students listen to Carl Sagan's essay "A pale blue dot." |
| Explore<br>"Plans and conducts investigations in which they observe, describe, and record data" | Students conduct a literature review.<br>Topics:<br>a. Animal Migration<br>b. Early human migratory patterns<br>c. Human migration into the United States<br>d. Current political rhetoric about immigration |
| Explain<br>"Explains using evidence from investigations" | Students engage and meet in uniform groups. |

| Sequence of events | Notes |
|---|---|
| Elaborate<br>"Makes conceptual connections between new and previous experiences" | Students meet in their mixed-topic groups and exchange information. |
| Evaluate<br>"Assesses progress by comparing current understanding with prior knowledge" | Students write a 500-word essay about what they learn about human migration and address the driving questions. |

## Engage

Students begin the assignment by priming their thinking by watching a short video like We Are Wanderers—Carl Sagan | A Profound Speech (Mudi Tracks, 2020), in which Sagan reads part of his 1997 book The Pale Blue Dot,

> *We were wanderers from the beginning. We knew every stand of tree for a hundred miles. When the fruits or nuts were ripe, we were there. We followed the herds in their annual migrations. We rejoiced in fresh meat. Through stealth, feint, ambush, and main-force assault, a few of us cooperating accomplished what many of us, each hunting alone, could not. We depended on one another. Making it on our own was as ludicrous to imagine as was settling down. . . . When the drought was prolonged, or when an unsettling chill lingered in the summer air, our group moved on—sometimes to unknown lands. . . . We could always begin again.*
>
> (p. 5)

Cited Works for video above

Mudi Tracks. (2020, March 5). We Are Wanderers—Carl Sagan | A Profound Speech [Video]. *YouTube*. https://youtu.be/NVuBV16Hh5o

## Explore

There were four major topics in the literature review. One student in each quartet was in charge of a specified topic, becoming an "expert" by exploring one of the following:

**Table 3.1** Potential data sources for the literature review portion of the migration lesson.

| | |
|---|---|
| **Animal migration:**<br><br>**NGSS: HS-LS4–2, HS-LS4–5, HS-LS2–8.**<br>**AP (EUs): EVO-1, SYI-3, IST-1.**<br>Nebel, S. (2010). Animal Migration. *Nature Education Knowledge*, 3(10), 77. www.nature.com/scitable/knowledge/library/animal-migration-13259533/<br>Parry, W. (2010). Why Do Animals Migrate? Retrieved November 30, 2020, from www.livescience.com/10235-animals-migrate.html<br>BBC Earth Unplugged. (2017). *Epic Animal Migrations*. https://youtu.be/EMshFxmNQhs | **Early Human migratory patterns:**<br><br>**NGSS: HS-LS4–6, HS-ESS3–1.**<br>**AP (EUs): EVO-1, SYI-3, IST-1.**<br>Love, C. (2019). *In Their Footsteps: Human Migration Out of Africa*.www.nationalgeographic.org/article/their-footsteps-human-migration-out-africa/12th-grade/<br>Gugliotta, G. (2008). *The Great Human Migration*. www.smithsonianmag.com/history/the-great-human-migration-13561/<br>Science Insider. (2015). *Map Shows How Humans Migrated Across the Globe*. https://youtu.be/CJdT6QcSbQ0 |
| **Migration into the USA:**<br><br>Roberts, S. (2009). Who's Coming to America. *New York Times*. https://archive.nytimes.com/www.nytimes.com/learning/teachers/featured_articles/20090413monday.html<br>Teaching Tolerance. (2011). Ten Myths About Immigration. www.tolerance.org/magazine/spring-2011/ten-myths-about-immigration<br>Business Insider. (2016). *Animated Map Shows History of Immigration to the US*. www.youtube.com/watch?v=Fe79i1mu-mc&ab_channel=BusinessInsider | **Trump administration immigration rhetoric:**<br><br>Arce, J. (2019). Trump's Anti-Immigrant Rhetoric Was Never About Legality—It Was About Our Brown Skin. *Time*. https://time.com/5645501/trump-anti-immigration-rhetoric-racism/<br>Fritze, J. (2019). Trump *Used Words Like "Invasion" and "Killer" to Discuss Immigrants at Rallies 500 Times: USA TODAY Analysis*. www.usatoday.com/story/news/politics/elections/2019/08/08/trump-immigrants-rhetoric-criticized-el-paso-dayton-shootings/1936742001/<br>Urrea, L. (2020). American Caudillo. *The Atlantic*.www.theatlantic.com/magazine/archive/2020/11/trump-american-caudillo/616467/ |

1. Animal migration: nonhuman animal migration including migratory patterns of whales and monarch butterflies in North America, as well as wildebeests in Africa.
2. Early human migratory patterns, namely the patterns of humans out of Africa around 65,000 years ago.
3. Human (historical) migratory patterns into the territory now comprising the United States, from those early migratory patterns of people that crossed around 20,000 years ago into what has happened over the last 500 years due to mass European migration and the enslavement of Africans.
4. Recent political rhetoric about immigration made by political figures.

Resources can be provided for each of the topic areas, but students were free to find additional information using their critical thinking skills to evaluate the validity of the sources of information. Some potential sources of information include those shown in Table 3.1.

### Explain

After each expert conducts their basic research, students will meet in uniform groups (e.g., all experts on animal migration) to ensure the students are able to exchange information and share their experiences before going back to meet with their mixed-topic working group.

### Elaborate

Now students meet in their mixed-topic group and exchange information about the four topics. Students compared the "reasons for, advantages, problems, facts" with the "reasons against, disadvantages, solutions, opinions," by using a t-chart graphic to organize their insights and record new information in a purposeful way. Once students determine the type of information that was to be recorded in each column, they add details based on prior knowledge, background information, additional research, and student comments during class discussion.

The focus questions driving group discussion include:

- "What does science actually have to say about migratory patterns of animals?"
- "What about human migratory patterns?"
- "What does history say about migration into the United States?"

### Evaluate

Students write their final 500-word essays using high-quality critical analysis that blended biology knowledge with analysis of modern rhetoric.

A single-point rubric can be used to assess the writing (see example in Activity 5).

# 4

# Lessons On Diversity and Taking Action

<table>
<tr><th colspan="2">Next Generation Science Standards</th></tr>
<tr><td>Science & Engineering Practices:<br>Asking Questions<br>Constructing Explanations &<br>Designing Solutions<br>Engaging in Argument from Evidence</td><td>Cross Cutting Concepts:<br>Stability & Change<br>Patterns<br>Cause & Effect: Mechanism &<br>Explanation</td></tr>
<tr><td colspan="2">Supported NGSS Standards:<br>HS-LS2–7: Human Impact Reduction Solution: Design, evaluate, and refine a solution for reducing the impacts of human activities on the environment and biodiversity. (Stability and Change)<br><br>HS-LS2–8: Social Interactions and Group Behavior: Evaluate the evidence for the role of group behavior on individual and species' chances to survive and reproduce. (Cause and Effect)<br><br>HS-LS4–6: Human Impact on Biodiversity Solution: Create or revise a simulation to test a solution to mitigate adverse impacts of human activity on biodiversity. (Cause and Effect)<br><br>HS-LS4–1: Evidence of Common Ancestry and Diversity: Communicate scientific information that common ancestry and biological evolution are supported by multiple lines of empirical evidence. (Patterns)</td></tr>
</table>

DOI: 10.4324/9781003409144-6

<table>
<tr><td colspan="2">Disciplinary Core Ideas:<br>LS4.C: Adaptation; LS4.D: Biodiversity and Humans; ETS1.B: Developing Possible Solutions; LS4.A: Evidence of Common Ancestry and Diversity</td></tr>
<tr><td colspan="2">AP Biology</td></tr>
<tr><td colspan="2">Essential knowledge:<br>EVO-1.F: Explain how humans can affect diversity within a population.<br>EVO-1.G: Explain the relationship between changes in the environment and evolutionary changes in the population<br>EVO-1.F.1: Through artificial selection, humans affect variation in other species.<br>EVO-1.G.1: Convergent evolution occurs when similar selective pressures result in similar phenotypic adaptations in different populations or species<br>SYI-3.B.1: Environmental factors influence gene expression and can lead to phenotypic plasticity. Phenotypic plasticity occurs when individuals with the same genotype exhibit different phenotypes in different environments.</td></tr>
<tr><td colspan="2">Vision and Change</td></tr>
<tr><td>Core Competencies:<br>1) the ability to apply the process of science; 4) the ability to tap into the interdisciplinary nature of science; 5) the ability to communicate and collaborate with other disciplines; and 6) the ability to understand relationships between science and society.</td><td>Core Concepts:<br>1) evolution (the diversity of life-forms that have evolved over time through mutations, selection and genetic change, 5) systems (the ways in which living things are interconnected and interact with one another).</td></tr>
</table>

As an introduction to the activities in these lessons, please check the overview videos of the authors, and other experts, as these videos provide an overview of the content and the philosophy behind the material. The video for this chapter can be found here: https://youtu.be/zr3np-YPPys

## Activity 6: Benefits of Diversity and Inclusion

### Overview

Students will learn about the immediate and long-term benefits of an inclusive, pluralistic multiracial democracy. Students will then look at how diversity can provide benefits to scientific fields and society in general. Students will read a total of three short articles about the importance of biological diversity and then reflect on their ideas about inclusion and diversity.

Suggested Timing: 45–60 minutes
Materials: Student notebook, internet access, or printed versions of the articles

**Driving Questions:** Why is diversity an asset in our society and our biological world? In which ways is biological diversity essential for our survival as a species?

### 5E Model of Activity 6

| Sequence of events | Notes |
|---|---|
| Engage<br>"Shows interest in the topic through curiosity and expression of wonderings" | Students read the Nature article "Collaboration: Strength in diversity," and write questions about the topics in the reading. |
| Explore<br>"Compares ideas with those of others" | Students read "When bringing in more diverse researchers broadens the scope of science" |
| Explain<br>"Explains using evidence from investigations" | Students create a small poster/display describing the reasons why science/society benefits from diversity. |
| Elaborate<br>"Makes conceptual connections between new and previous experiences" | Students will write a 35-word response. |
| Evaluate<br>"Assesses progress by comparing current understanding with prior knowledge" | Students provide a clearly-stated rationale for why Hitler was both morally and biologically incorrect about creating a "master race." |

## Engage

Students will begin by reading the Nature article "Collaboration: Strength in diversity," 2014.

www.nature.com/news/collaboration-strength-in-diversity-1.15912. After this short reading, students will generate questions and write them in a public place where others can read them. These two questions can be used to guide students: What questions were raised while reading? Where can we go from here?

## Explore/Explain

Then, students read the article "When bringing in more diverse researchers broadens the scope of science" www.anthropocenemagazine.org/2020/09/when-bringing-in-more-diverse-researchers-broadens-the-scope-of-science/?fbclid=IwAR24uVj_cspPWgEeL8P6K2CeybZW55uQVsASIsDDp7NxHaV_I1K5AQYQ-uk.

Then, students read the article "Social Justice isn't just for super heroes," www.happiness.com/en/magazine/inspiration-spirituality/social-justice-benefits-everyone/

Finally, students work in pairs to create a poster/diagram that relates why science/society benefits from human diversity.

## Elaborate

Genetic diversity is essential for the survival of biological species in an ever-changing environment. As a whole class, you can review the information from the Pasteur Institute on the benefits of genetic diversity: www.pasteur.fr/en/research-journal/reports/human-evolutionary-genetics-benefits-genetic-diversity or watch this short video directly on YouTube (in French with English subtitles): https://youtu.be/LypUxFiJxmA.

As a summary of this information, students will write a short descriptor limited to 280 characters (or 25 words)

*Teacher Tip:*

*Genetic diversity provides populations a way to adapt to changing environments. Greater variation increases the chances that some individuals will possess variations of alleles that are suited for changing environments. Additionally, hybrid offspring with heterotic traits can have improved or increased function as a result of mixing the genetic contributions of its parents, resulting in hybrid vigor or heterosis.*

## Evaluation

Students will respond to the prompt below for homework. The responses should be logical and evidence for claims should be provided:

> Given what you know about society and biology, provide a clearly-stated rationale for why Hitler was both morally and biologically incorrect about creating a "master race." Think about his concept of race and why it is incorrect. Additionally, consider the benefits of diversity in biological systems, and the benefits of diversity in society.
> Student work can be submitted in a variety of ways including a poster, a brochure, a short video, an audio clip (radio announcement), or direct writing.
> Student products can be evaluated using a single-point rubric.

| Areas that need work | Criteria | Things that were amazing |
|---|---|---|
| | Your ideas, content, organization and word choices were right on! Your voice and/or use of graphics provided the right amount of appeal. | |

## Activity 7: Making Connections/Reflection

### Overview

By this point, students have explored race and racism and can now begin to design procedures that could educate people about the impact of racism.

These procedures that could teach people about the truth and scientifically validated ideas about human grouping, need to be based on scientific evidence. Therefore, this part will provide opportunities for students to think about and record their individual and collective learning as they progress through the activities.

Suggested Timing: 45–60 minutes
Materials: Notebooks, internet access

**Driving Questions:** How can what you are learning in your biology class support solutions that improve [reduce racism] in your local and global community?

### 5E Model of Activity

| Sequence of events | Notes |
|---|---|
| Engage<br>"Shows interest in the topic through curiosity and expression of wonderings" | Students reflect and create a shared list of answers to the question, "What are the social impacts of racism?" |
| Explore<br>"Compares ideas with those of others" | Students will research current programs/entities addressing racism and inequity. |
| Explain<br>"Explains using evidence from investigations" | Students will share their findings as guided by the questions posted around the room. |
| Elaborate<br>"Makes conceptual connections between new and previous experiences" | Students will run a campaign to inform others about the impact of racism on our society. |
| Evaluate<br>"Assesses progress by comparing current understanding with prior knowledge" | Students will create a PSA about this topic. |

## Engage

Students begin by creating a shared list of answers to the question, "What are the social impacts of racism?" This can be done by using sticky notes or electronically by using a shared document.

## Explore

Students should now research programs/entities that address social inequities based on racism.

*Teacher Tip: There are several organizations working on issues related to issues currently addressing racism. These organizations should be split among groups of students and each group will be responsible for sharing basic information about their assigned organization to the rest of the class:*

*Southern Poverty Law Center. (2022, December 14). www.splcenter.org/*
*American Civil Liberties Union. (2022). www.aclu.org/*
*Equal Justice Initiative. (2022, December 16). https://eji.org/*
*Racial Equity Resource Guide. (2022). www.racialequityresourceguide.org/*
*NAACP Legal Defense Fund (2022). www.naacpldf.org/*

## Explain

As students investigate existing programs addressing racism, they should answer the questions posted around the room: 1. What did you feel these programs do well, and what did you feel they could do better? 2. Who should be responsible for improving racism locally and globally? 3. What role do you think you could play in addressing racism locally and globally? 4. Based on what you learned about racism, and the actions others are already taking, what are five areas of need that you could address?

## Elaborate

Students will summarize their learning to educate their classmates on the issue they have identified and investigated related to the issue of racism. A choice of formats is provided and includes class presentations, posters, blog posts, etc. Summaries may be supported by multimedia and/or print materials that synthesize and analyze the topic and issue on local and global levels. Options include: A) Create and deliver an educational workshop to raise awareness about the topic and its local impact with a strong call to action that leads to enacting change. These workshops can be geared toward younger students, teachers, elected officials, and/or the community at large. B) Volunteer at an organization that works for global issues—many organizations offer ways to get involved on their websites and in their offices. C) Create a campaign writing letters to the United Nations, government bodies, and other leaders to ask for added resources on the issues.

*Teacher Tip*

*As a society, we are all only as strong as the weakest among us. As Dr. Martin Luther King Jr. wrote, "Injustice anywhere is a threat to justice everywhere. Whatever affects one directly, affects all indirectly." There is great hope in our students and the intent of this module is not to leave them feeling overwhelmed by the horrific mistakes of the past, we must provide them with the hopefulness of a better future. There are a number of actions that students can take to improve race-based injustices in their local communities and around the world right now.*

## Evaluate

As an evaluative reflective component, students can create a video demonstrating the impact of these lessons. This video project can be considered a Public Service Announcement (PSA). Racism, unequal treatment, and opportunities cause harm and it goes against the principles of equality and respect, which are fundamental to a fair and just society.

To assess student work, the following rubric can be utilized:

| Areas that need work | Criteria | Things that were amazing |
|---|---|---|
| | Message is clear and effectively communicates the intended information. | |
| | Production values (e.g., sound quality, visuals) are high and enhance the PSA. | |
| | PSA is engaging and holds the audience's attention. | |
| | PSA is appropriate for the target audience. | |
| | PSA inspires the audience to take action or change their behavior. | |

Lastly, to help students think critically about their work and reflect on their experiences, the following set of questions are meant to help students to identify their strengths and areas for improvement and encourages them to think about how they can apply what they have learned to future projects and assignments.

Thinking about the set of lessons you just completed, please answer the following questions:

What did you enjoy about these lessons? Why?
What challenges did you face during these lessons? How did you overcome them?
What did you learn from these lessons? How will this learning impact your future work?
What feedback did you receive on the assignments? How will you use this feedback to improve your work in the future?
Is there anything you would do differently if you were to complete a similar project or assignment in the future? Why?
What actions can you take to help reduce racism and the negative impact it has on society?
How can you educate others about the way that humans are grouped and related?

## Additional Content

## Additional Activity: Understanding Variation in Human Skin Color

### Overview

Students can now dig deeper into the biology of skin pigmentation patterns. Using the HHMI BioInteractive activity "Understanding Variation in Human Skin Color": www.biointeractive.org/classroom-resources/understanding-variation-human-skin-color. Students will complete the HHMI activity and apply models and mathematical thinking to explain how genomic variation and human ancestry can explain differences in the polygenic trait of skin color.

Suggested Timing: 45–60 minutes
Materials: Student notebook, internet access (for HHMI website)

**Guiding Questions:** What were the environmental pressures that led to the different skin pigmentation patterns that we observe in humans today?

How can evolutionary theory account for the current phenotypes we see in human skin color pigmentation patterns? (see Chapter 7—The theory of evolution by natural selection).

*Teacher Tip: This activity can be done asynchronously and then come together to discuss the information presented.*

### Check for Understanding

Collectively (as a class discussion), students can conclude and summarize the main learning objectives met by this activity.

## Potential Assessment Questions

These questions connect biology with the material covered in this module and could be used as part of a summative assessment.

### Multiple-Choice Question

Melanin is a dark pigment in the skin that can absorb most ultraviolet (UV) radiation. Different populations of humans have different average amounts of melanin in the skin. Populations of humans with the darkest skin colorations have typically lived nearest to the equator.

*Which of the following pieces of evidence best supports the claim that patterns of skin pigmentation in human populations are the result of natural selection?* [Key: B]*

A) Humans can develop skin cancer regardless of the amount of skin pigmentation.
B) Ultraviolet (UV) radiation is most intense near the equator and least intense toward the poles.
C) Humans migrated out of Africa around 70,000 years ago and now live throughout the globe.
D) Individuals can produce greater amounts of melanin based on ultraviolet (UV) exposure.

### Free-Response Question

Data collected over many years by NASA show that UV radiation is most intense near the equator and least intense toward the poles. Sunlight is essential to our healthy development and our immune systems, because sun-exposed skin produces Vitamin D. However, overexposure to sunlight can damage cells by disrupting DNA and breaking down folate (an essential B-vitamin needed to make red/white blood cells in the bone marrow, convert carbohydrates into energy, and produce DNA and RNA).

Melanin is the skin pigment that protects the skin from DNA damage and circulating folate from being broken down by UV radiation. Populations with darker skin color (or more eumelanin) are found in regions with more intense UV radiation. Thus, populations found in equatorial areas have the darkest skin (most eumelanin) and populations at higher latitudes (with less UVB radiation) will have lighter skin (least eumelanin).

*Part A:* Describe why individuals that migrated out of Africa around 70,000 years ago were most likely phenotypically darkly pigmented.

*Part B:* Levels of vitamin D and folate are directly impacted by UV light. Describe the potential impact of using a tanning booth (which exposes skin to UV light) on the blood levels of these two vitamins (based on questions from HHMI).

*Part C:* Predict the effect of the recent human migration on the circulating levels of folate and vitamin D. Discuss the concerns of a lightly pigmented human around the equatorial region (with high sunlight exposure) and of a darkly pigmented human in regions where little sunlight is experienced.

*Part D:* Evaluate why the traditional way of grouping humans by races is morally and scientifically incorrect. Describe how these racial groups have been utilized to create a hierarchy of human groups and oppress some people.

*Part E:* Given our knowledge of the benefits for genetic biodiversity, justify the assertion that as a society we should be more inclusive and supportive of people who have historically been oppressed.

# Part III

## Racial Justice; Pedagogy, and Evolutionary Biology

# 5

# The Meaning of Racially Based, Social Injustice

*On June 17th, 2020, I, along with several students, parents, and board members from Central Falls High School (RI) where I teach, stood on the steps of the Rhode Island State House and addressed a crowd of approximately 150 people. We came together to protest an unjust school budget that led to the firing of 42 teachers, including all the art and physical education teachers, the librarians, and the nurses (Cullinane, 2020). This is some of what I said:*

> *"I've been here before. I was a 15-year-old who didn't know much about budgeting or funding, but I had a clear understanding of fairness when a couple of my teachers took me to this statehouse to speak on behalf of my school's limited budget and its detrimental results. I spoke about my awareness that we in Central Falls—lacking books, computers, a reasonable music program, and saddled with a decrepit building—knew we were slowly being starved.*
>
> *"I can't believe that 28 years later, I am back here speaking about the same issues. The decrepit building only further battered by time, the music program actually cancelled, the materials still limited—the state of RI could have made Central Falls an innovative hub for education in the state. Instead, it has opted to underfund the schools and oppress the children by offering less than is needed.*
>
> *". . . can you look at the face of any of my students and tell them they are not worth art, physical education, a library, a nurse, a music class? We want opportunities to live our best lives. We know that education can develop the mind to think critically and that kind of freedom in and of itself is invaluable, but education can also lead to higher incomes and higher incomes lead to more choices. So, I want everyone here to say it: EDUCATION IS FREEDOM."*

DOI: 10.4324/9781003409144-8

*I stood on those steps thinking about the fact that no other school system in RI needed a rally to ask for basic funding. So many times, the students from the poorest district in RI, serving a student body with 12.8% Black and 72% Hispanic students (Civil Rights Data Collection: Central Falls, 2020), had been traumatized and re-traumatized as they saw their teachers being laid off, their school buildings crumbling, and their options increasingly limited. No other school system in RI has eliminated music, art, and physical education, the librarians, and the school nurses. The signs that some people carried in the crowd were indicative of an ugly reality and we were attempting to right it:*

> *"Public education is an investment, not an expense! Proper funding for Central Falls!"*
>
> *"Central Falls students desperately need equitable funding to learn and grow!"*
>
> *"Don't deny our children a future they deserve."*
>
> *"A zip code should not dictate educational opportunities."*
>
> *"Why should brown kids get less?"*
>
> *"You are lucky we're asking for equity, not revenge!"*

Our need for social and racial justice focused our frustrations into an organized event where people could hear the policies and practices that oppressed the students of Central Falls. In fact, the poorest school in Rhode Island and its surrounding wealthier communities are a microcosm of the larger society, where some people have access to opportunities, while others are denied these. Adequate funding is fundamentally needed to provide students with access to a high-quality education, which is essential for social mobility and breaking the cycle of poverty. However, as is the case with Central Falls, many schools in BIPOC, low-income communities receive significantly less funding than their more affluent counterparts, leading to a lack of resources and opportunities for students. This further perpetuates a cycle of educational inequality and limits the ability of marginalized students to succeed. In turn, this inequitable funding has long-term effects on society, as underfunded schools tend to produce less educated, less skilled graduates who are more likely to be unemployed or underemployed, more likely to be dependent on welfare, more likely to have poor health, and more likely to turn to criminal activity to survive. Therefore, addressing school funding disparities and ensuring that all students have access to adequate resources is an essential aspect of achieving social justice.

Although injustices can be traced to many different sources, here we focus on injustices based on race and ethnicity, which are closely linked to income and wealth.

## What is Racial Injustice?

In this chapter, we introduce the second part of our curricular duet, racial justice, and we will do this by way of exploring the more pertinent topic at hand, racial *in*justice. We trace the history of racial injustice in the United States, presenting evolving concepts of what *justice* would look like along the way. With that introduction, then we turn to the grim legacy of racialized social injustices practiced in the United States; not simply when Reconstruction failed in the hands of President Andrew Johnson—a southern former congressman (Tennessee) far less committed to equal rights than his predecessor, Abraham Lincoln—but relentlessly and creatively carried out until the present day. The timeliness of our inquiry cries out for a way to transmit the concept of racial injustice, including its history and its illegal and immoral subversion of American society, to future generations.

We start with the broad umbrella of social justice. Modern concepts of social justice effectively begin with European enlightenment philosophers. Scottish philosopher David Hume (1711–1776) reflected on the attributes and conditions necessary to produce an ideal moral agent and social justice system in his *A Treatise of Human Nature*, originally published in 1740 (2000). Hume posited that individuals would cooperate with others whom they believe will not break commonly accepted and enacted rules of behavior, creating the necessity for social norms. Similarly, German philosopher Immanuel Kant (1724–1804) described people as free moral agents, endowed with the capacity to make decisions independently of external causal forces and their own desires, committing themselves to what they recognize as universal, rational constraints on their own action (Rosen, 1993).

In 1762, Swiss philosopher Jean-Jacques Rousseau's (1712–1778) *The Social Contract* (1762), promoted the notion that society needs a contract as a way to address the unequal and unjust systems. In *The Social Contract*, Rousseau argued against the idea that monarchs were divinely empowered to legislate and instead affirmed that only the people, who are sovereign, have that right to govern by and for themselves. Hence, in Rousseau's telling, individuals should be seen not as subordinate subjects but as equal members of societies. Rousseau's concepts of the commonwealth inspired political reforms and revolutions,[1] including the American Revolution, and provided its foundational philosophy (Damrosch, 2012).

The concept of social justice as the societal variant of justice gained popular currency during the Age of Enlightenment in the early 1600s and can be traced through the philosophy of English-born American philosopher Thomas Paine (1736–1809) who, in his last pamphlet, *Agrarian Justice* (1797), opposed agrarian law and monopoly by proposing the idea that the basic needs of all humanity must be provided for by those with property, since to him, the earth was the common property of the human race. Moreover, the term "social justice" appears in foundational documents such as *The Federalist Papers, No. 7* (1787) when Alexander Hamilton (1757–1804) argued for the importance of the union of states to the well-being of Americans by describing the likelihood of retaliation of one state with another due to legislation, "would chastise such atrocious breaches of moral obligation and social justice" (Hamilton, 1787/2008).

In the United States, social justice today is dominated by struggles for racial justice, which can be traced back to the 18th-century slave revolts and "Maroons" who escaped slavery to create independent communities on the outskirts of slave societies, and notions of justice for the Indigenous struggles (Robinson, 2000). Moreover, modern ideas of social justice also emerged out of the worsening inequities of the European working class by the capitalist mode of production in the mid-19th century, and one of the central figures of this thinking was Karl Marx (1818–1883). Marx proposed the idea that society should provide individuals what they need to fully develop their talents, and that each individual in return should provide society with work for the good of society (McCarthy, 1990). Since the time of Marx's writings, the concepts of social justice and the related concept of human rights maintain their relevancy in the face of unbridled industrialization, unjust policies, and oppressive actions.

Although there have been many different conceptions of social justice, including radical egalitarianism, welfare-based, and libertarianism, their theoretical positions converge in a composite idea of fairness/equity/equality (Bales, 2018). While many of these ideas focus on economic factors and the equitable distribution of material resources, social justice is more than just a way of understanding material apportionment. Marx affirmed this when he proposed that the ultimate object for society is the full realization of an individual's humanity and gaining true freedom which would return to people the agency to realize their liberation from the exploitative material and ideological restraints. Marx saw these as limiting peoples' humanity (McCarthy, 1990). Originally written in 1846, but first published in 1932, in *The German Ideology* Marx wrote: "people cannot be liberated as long as they are unable to obtain food and drink, housing and clothing in adequate quality and quantity. 'Liberation' is a historical and not a mental act, and it is brought about by historical conditions" (p. 44). Social justice can be seen as transcending economic wealth by incorporating political, cultural, religious, and sexual freedoms, and liberating humanity from ideological constraints.

The ideas of Hume and German philosophers Kant and Marx, were built upon and articulated by 20th century thinkers such as Sir Isaiah Berlin (1909–1997), and Americans Thomas Nagel (1937–), John Rawls (1921–2002; see below), and Robert Nozick (1938–2002). These cotemporary philosophers discussed concepts of freedom, equality, and rights, as these pertain to justice and democracy.

Social justice ideals were reaffirmed with the creation of the United Nations in 1945 and its exploration of income-related inequalities, including ownership of capital and assets, and access to services and social benefits.[2] The modern ideals of social justice were then codified in a number of official statements and guidelines, including the United Nations' (1948) Universal Declaration of Human Rights (UDHR). Formally recognizing "the equal and inalienable rights of all members of the human family," the UDHR has served as a benchmark of human rights. This publication was a recognition that social justice interests would now become a matter of international importance because we currently live in a closely interdependent world where injustices impact all peoples (United Nations, 2006). This sense of shared responsibility was affirmed in the more recent document *Social Justice in an Open World: The Role of the United Nations (2006)*, which connects human rights to concepts of social and economic equity by identifying three critical domains of equality and equity, "equality of rights, equality of opportunities, and equity in living conditions" (pp. 15–16).

One of the most influential and preeminent theoreticians of fairness and social justice in the late 20th century was American political philosopher John Rawls, whose ideas about fair distribution and political freedom hinge on the existence of basic freedoms and equality of opportunity. In 1971, he proposed a society of free citizens that hold equal basic rights and cooperate in an egalitarian economic system (Rawls, 1999). He suggested that societies be structured in a way where the greatest possible amount of liberty is given to their members, as long as the liberty of any one member does not infringe upon that of any other member. He also stated that there should not be social and economic inequalities, with the exception of the situation where the worst off are better off than they might be under an equal distribution. Even then, if such an inequality is created, this inequality should not hinder those without resources from occupying positions of power, like public office. Rawls' 1971 fairness concept of shared equality, in a conscious echo of Rousseau, employed a social contract where people "could expect the willing cooperation of others when some workable scheme is a necessary condition of the welfare of all" (Rawls, 1999, p. 15). More recently, Cramme and Diamond (2009), define social justice as the fair and just distribution of rights, opportunities, and resources within a society—a more equitable distribution of opportunities and life chances than the one currently available to most people around the world.

Our call to action against the injustices in our world thus arises from a history of injustice that fails to attain the ideals set out by United Nations (1948, 2006) and by social philosophers ranging from the 17th to the 21st centuries.

With that introduction to the general concept of social justice, we now turn our attention to *injustice*, and to the particular case of racialized social injustice, or racial injustice.

It will be noted that in all cases cited above, the assumption is that social justice applies to all humans. Therefore, the ideal of social justice must apply to humans of all colors, unless the argument is made that persons of color, which for people of European ancestry has historically meant non-whites, are not fully human. As we shall see, this argument has been made (and continues to be made) explicitly as well as implicitly, leaving persons of color without the same "inalienable rights" to social justice as white people (see Chapters 8–10). And as we noted above, people with such viewpoints have generally resorted to science[3] as the ultimate justification for their views. Specious science has been invoked to justify these racist viewpoints, but ultimately, such science fails the most basic scientific tests (Chapter 10).

## Racially Motivated Injustices

Injustices can be categorized as *personal*, or *systemic*, but often may be combinations of both. *Personal injustices* are generated by prejudices in which an individual's decisions are the main source of biased and unfair treatment of others, while *systemic injustices* are those created by the laws, regulations, and social contracts that create oppression. Both personal and systemic injustices are associated with the racial oppression seen in the United States. For instance, racial segregation can be the result of oppressive actions deemed to be "*de jure*" (of law) segregation, which are actions by the state to enforce racial segregation (e.g., the "Jim Crow" laws of the southern states, the remnants of which endured until the 1960s), or segregation can be "*de facto*" (of fact), which is accomplished by factors outside of systemic and legal constraints. While here we have segregated *de facto* and *de jure* discrimination, these two categories are in practice generally mutually reinforcing.[4]

The systemic forces that feed social injustices may be based on many factors; however, those based on race are manifested by limiting access to quality healthcare, access to information, the right to essential nutrition, housing, work, education, incarceration rates, and restricting participation in the democratic process. Prilleltensky and Gonick (1996), describe social injustice as large-scale enactments of inequitable practices that can be understood from the macro-level of global economic and political structures to the micro-level of internalized psychological images of inferiority.

Some people would affirm that income differences, educational attainment, private discrimination of real estate agents, banks and other institutions can be described as *de facto* segregation—the results of accidents or individual choices. While we would deny neither the reality nor the perniciousness of *de facto* segregation, the idea that it is the dominant cause of inequities in the United States today, a myth that has enjoyed widespread currency across the political spectrum, obscures the truth that segregation has been imposed and maintained by racially explicit federal, state, and local policies, without which private actions of prejudice or discrimination would have been ineffectual. Only by recognizing the impacts of governmentally sponsored systems of segregation (*de jure* inequities), can we begin to remedy them by developing equally effective government actions to reverse them (Rothstein, 2017).

Here we present some major themes in racially motivated discrimination and systemic injustices, including disparities in housing, material wealth, prisons, schools, and mortality/health. These categories are not mutually exclusive, interact at a variety of levels, and are the result of both personal and systemic forces.

**Housing**. A key indicator of racialized social injustice comes in the form of residential segregation. In the United States white people tend not to live in the same neighborhoods as people of color. The data show that a typical white person lives in a neighborhood that is 75% white and 8% Black, while a typical Black person lives in a neighborhood that is 35% white and 45% Black (Logan & Stults, 2011). Yet, where people live surely matters for their short- and long-term social and economic well-being. In the United States, there are enclaves of low-income neighborhoods that lack the resources, amenities, and the political and market power that most affluent communities have.

According to the U.S. Department of Housing and Urban Development (2020), on a single night in January 2019, 567,715 people—about 17 of every 10,000 people in the United States—were experiencing homelessness across the United States. This metric disproportionately affects Blacks, who, despite the fact that they constitute 13.4% of the overall US population, constitute 39.8% of all homeless people. By contrast, Latinx people constitute 18.5% of the overall US population, but constitute 23% of all homeless people (U.S. Census Bureau, 2020a).

Shockingly for those who imagined that American racism was only *de facto*, racially segregated neighborhoods are the physical realization of plans, policies, and practices, carried out at federal level, state, and local levels, that have systematically denied equal opportunity to BIPOC populations. The most infamous (and arguably far-reaching) systemic racism was "redlining." This effectively began in the mid-1930s, when the Federal Housing Administration (FHA) advanced segregation efforts by "redlining," or demarking (with red borders on maps) locations where it refused to insure mortgages in and near Black neighborhoods, driving down home values in such neighborhoods and encouraging an exodus away from them by (generally white) people who saw their investments faltering. The FHA's *Underwriting Manual* (1936) specified its intent in unambiguous terms:

> *Usually the protection from adverse influences afforded by these means includes prevention of the infiltration of business and industrial uses, lower class occupancy, and inharmonious racial groups.*
>
> (paragraph 229)

> *Areas surrounding a location are investigated to determine whether incompatible racial and social groups are present, for the purpose of making a prediction regarding the probability of the location being invaded by such groups. If a neighborhood is to retain stability, it is necessary that properties shall continue to be occupied by the same social and racial classes. A change in social or racial occupancy generally contributes to instability and a decline in values.*
>
> (paragraph 233)

> *Schools should be appropriate to the needs of the new community and they should not be attended in large numbers by inharmonious racial groups.*
>
> (paragraph 289)

> *The restrictions should be recorded with the deed and should run for a period of at least twenty years. Recommended restrictions include the following . . . (g) Prohibition of the occupancy of properties except by the race for which they were intended.*
>
> (paragraph 284; our italics)

Based on such language in the FHA manual as well as his own investigation, Rothstein (2017) concluded that as part of this racist program, the federal government subsidized construction for white homeowners with the stipulation that the homes would not be sold to non-white persons, since it judged mixed (white and non-white) neighborhoods as "inharmonious," driving down home values and creating detrimental conditions for white children (see Rothstein, 2017).

An extraordinary expansion in home ownership was driven by the Servicemen's Readjustment Act (1944, the "GI Bill"), Although the GI Bill was (and is) race-neutral, state-controlled push-backs ensured that black veterans were kept from reaping its full benefits and in fact, the GI Bill increased racial wealth disparities (Darity & Mullen, 2020). Segregated schools, vocational training programs, and redlining kept the 1.2 million Black soldiers (Baker, 2016) that fought for the United States during World War II from enjoying GI Bill benefits (Blakemore, 2019). Black veterans were discriminated against, assaulted, and even lynched in the years following World War II (Equal Justice Initiative, 2017). Home ownership had been made accessible to white, middle-class, lower-middle-class families since 1944, while Blacks and other people of color were excluded from suburban communities and pushed instead into urban housing projects.

As part of H.S. Truman's "Fair Deal," the FHA, acknowledging the inequities of segregation, established the 1949 "American Housing Act" program explicitly aimed to eliminate urban slums. In the end, however, redlining had effectively made it nearly impossible for Black families to move into the white-dominated suburbs. Black families were displaced from urban housing, which was replaced with either unaffordable alternative housing, or public works that were still segregated (Rusk, 1999).

Despite the fact that more than a century's worth of discriminatory policies built the physically segregated America of today, governmental policies that resulted in residential racial segregation are clear violations of the Constitution and Bill of Rights: after the passing of the Thirteenth (1865) and Fifteenth (1870) Amendments, in which Blacks stopped being the legal property of white slave owners, and their citizenship and right to vote were fully recognized. The Fifth Amendment (1794) promoted equitable treatment of citizens, who could not "be deprived of life, liberty, or property, without due process of law," while the Fourteenth Amendment (1866–1870) affirmed that

> No State shall make or enforce any law which shall abridge the privileges or immunities of citizens of the United States; nor shall any State deprive any person of life, liberty, or property, without due process of law; nor deny to any person within its jurisdiction the equal protection of the laws.

Taken together, all of these Amendments extend equal protection in home buying, as well as protection against other segregated practices to *all* Americans (which unequivocally includes BIPOC).

**Material Wealth.** As was the case with housing, racial disparities in wealth also have a connection to the GI Bill. When the original bill expired in 1956, almost 8 million World War II veterans had received education or training benefits, and 4.3 million home loans worth $33 billion had been distributed (Blakemore, 2019). However, most Black veterans were left behind, since they faced limited options due to lack of formal secondary education, and often, outright denial in their pursuit for educational advancement (Thompson, 2019). As a result, employment, college attendance and wealth surged for whites but not for BIPOC, widening the racial wealth disparities.

Wealth, the total value of a family's financial resources minus all debts, offers a key to understanding racial stratification in the United States. Using wealth as a metric, the net worth of typical white families is $171,000 compared to $17,150 for black families and $20,720 for Latinx families (Tax Policy Center, 2019).

Because in America, home and land ownership are perhaps the most significant repositories of private wealth, these too reflect racially-based disparities. According to recent Census Bureau data (2020a), the percentage of non-Hispanic white people who own their home is 73.3%, compared with 42.1% for Blacks and 47.5% for Latinx. This means that a much greater percentage of whites have wealth, relative to non-whites; here, then, redlining and other oppressive measures effectively manipulated wealth distribution in the nation.

In the United States in 1989, the top 10% of households controlled around 60% of the total wealth. This trend has been further cemented since by 2019 those top 10% of households now control around 70% of the total wealth (FRED. Federal Reserve Bank of St. Louis, 2019). Importantly, this wealth distribution has an inherent intergenerational mobility, that is to say, that children of wealthy people are likely to remain wealthy due to a variety of factors including better health care, better schooling, and more cultural capital (Reeves, 2017). This unequal pattern of wealth apportionment is also reflected in the difference between CEO compensation and that of the average worker in manufacturing or production. CEO compensation has grown 940% since 1978, while typical worker salaries rose only 12% during the same time period (Mishel & Wolfe, 2019). As of the time of this writing, out of all Fortune 500 (500 of the largest US companies ranked by total revenues for their respective fiscal years) CEOs, there are a total of three Black CEOs, accounting for less than 1% (Davis, 2020), and 11 Latinx CEOs (~2%; Reiss, 2018). The growing separation between the top 20% of wealth owners and the rest of society exacerbates the race gaps in access to some of the mechanisms of class mobility. For example, inheritance of economic wealth, educational attainment, family formation, geography, and health and life expectancy, are ways in which class and race divisions amplify each other (Reeves, 2017). The inheritance of social class, or what Reeves (2017) calls "opportunity hoarding" by the wealthy and mostly white, entail the systemic limiting of access to wealth-building opportunities.

Given the pernicious history of redlining, it is not surprising that there is an over-representation of people of color in high-poverty areas in the United States. A low-income Black person is more than three times as likely to live in a neighborhood with a poverty rate of 40% or more than a white person is, and a low-income Latinx person is more than twice as likely to live in such a neighborhood (Jargowsky, 2015). These statistics reflect the racial residential segregation and racialized concentrated poverty that plague our society.

Even when doing the same job, BIPOC earn less than their white counterparts. Using data from a sample of 1.8 million employees surveyed between January 2017 and February 2019, on average, black men earned 87 cents for every dollar a white man earned, while Latinx workers earned 91 cents for every dollar earned by white men (Gruver, 2019).

All of this is happening in the country with the largest economy in the world, which has a nominal Gross Domestic Product (GDP) of $21.43 trillion and a GDP per capita of more than $65,000. In fact, the United States has been the world's biggest economy since the late 19th century (International Monetary Fund, 2020), and 2019 Census documentation shows racial disparities in real median incomes of non-Hispanic whites which are significantly higher ($70,642), than Black ($41,361), and Latinx ($51,450) households (U.S. Census Bureau, 2019a). Additionally, in 2016, approximately 18% of school-age children were in families living in poverty and the percentage of children under the age of 18 in families living in poverty was higher for Black children and Latinx children (31 and 26%, respectively), than for white and Asian children (10% each; National Center for Education Statistics, 2017).

Another key indicator of racial wealth disparity shows up in bankruptcy data. Using a national dataset of bankruptcy filings, Kiel and Fresques (2017) documented the scope of racial disparities in bankruptcy filings and outcomes, finding large, pervasive disparities that remained even after factoring for income, assets, and the court districts where debtors lived. Their analysis shows that for Black people who file for bankruptcy, the odds of having their cases dismissed were more than twice as high as those of white debtors (Kiel & Fresques, 2017).

As we have seen, racially correlated wealth disparities are the result of unjust policies and personal prejudices that have enabled the systemic oppression of BIPOC and resulted in detrimental cycles of poverty at every institutional level and turn. Other systemic injustices are recognizable in our legal system, schools, and health.

**Justice system/Prisons.** Significant evidence for racialized social injustice in the United States includes incarceration, which has infamously grown dramatically since the 1970s and made the United States the country with the largest total number of prisoners in 2020 with the highest rates in the world, at 655 prisoners per 100,000 of the national population (Walmsley, 2019). This rise in incarceration has been especially prominent among young Black males and high-school dropouts. Prison data are among the most egregious examples of underlying racist behavior in the judicial system; for example, the percentage of men aged twenty to thirty-four who are in prison and are high school dropouts is, among Black men, 37%, compared to 12% of white men with the same education level (Western & Pettit, 2010). Moreover, data show that black male federal defendants receive longer sentences [approximately 10%] than whites arrested for the same offenses and with the same prior records (Rehavi & Starr, 2012). Similarly, the women's incarcerated population is now eight times higher than it was in 1980 and prison data show that in 2017; the imprisonment rate for Black women (92 per 100,000) was close to twice that for white women (49 per 100,000); while Latinx women were imprisoned at 1.3 times the rate of white women (67 vs. 49 per 100,000) (Bronson & Carson, 2019).

For young Black and Latinx men, police use of force is among the leading causes of death[5] (Edwards et al., 2019). Disproportionally, compared to all other industrialized countries, one in every 1,000 black men can expect to be killed by police. In fact, the probability of being black, unarmed, and shot by police is about 3.5 times the probability of being white, unarmed, and shot by police (Ross, 2015). Edwards et al. (2019) reported that between the ages of 25 and 29, black men are killed by police at a rate between 2.8 and 4.1 per 100,000, Latinx men at a rate between 1.4 and 2.2 per 100,000, while white men at a rate between 0.9 and 1.4 per 100,000—and these inequalities in risk persist throughout life. These and other acts of physical violence against racially minoritized people have been a persistent problem in our society.

**Schools**. In the United States, it is accepted that we have separate and unequal schools segregated by race (among some other criteria). Yet, as a nation, we say that we expect our schools to create equal outcomes for students who live in different communities, even if students live in poverty, have limited health care access, their schools have unqualified or inexperienced teachers, and those schools are underfunded. The data show that high-school graduation rates are highest among whites (89%) and lowest among Native Americans (72%), Blacks (78%) and Latinx (80%). Similarly, college graduation rates are lowest among Blacks (26.3%) and Latinx (18.8%) and highest among Whites (40.1%) (National Center for Education Statistics, 2019).

The Federal government's attempt to help schools improve student outcomes has been largely a punitive approach where schools serving students with the greatest needs are punished for not achieving positive changes in standardized tests. There are stubborn race gaps on standardized tests scores, which provide a snapshot into the astonishing extent of racial inequality in our education system. Although standardized tests are often portrayed as mechanisms for meritocracy, test scores reflect accumulated advantages and disadvantages in the lives of those taking the tests. Hence, race gaps on SAT and ACT hold up a mirror to racial inequities in society rather than reflecting academic abilities (Reeves & Halikias, 2017). In terms of school funding, the financial punishments that result when schools serving poor children fail to improve standardized test scores encourages the teachers and administrators to move elsewhere, leaving behind a revolving door where students with the greatest educational needs commonly encounter new and/or inexperienced teachers and administrators (Orfield, 2008).

While facile readings of figures such as those cited above, and others like them, have been used historically to suggest unequal academic capability among different races (and more recently by Herrnstein & Murray, 1994; Richwine, 2009; Lynn, 2006), a substantial and broadly accepted literature has demonstrated that these differences derive from a complex suite of issues ultimately stemming from a wide variety of racist sources (e.g. Jencks & Phillips, 2016; Ladson-Billings, 2006; Gillborn et al., 2017; Clark, 2014; Darling-Hammond, 2014). Educational outcomes for BIPOC children are much more likely to be a function of their unequitable access to educational resources, including skilled teachers, class sizes, and quality curriculum, than they are a function of racial differences. In fact, the US educational system is one of the most unequal in the developed world and, based on their social status, students routinely receive dramatically disparate learning opportunities (Darling-Hammond, 2016).

Some oppressive practices can be linked to *de facto* biases that teachers bring to the classroom. For example, some teachers tend to perceive someone with darker skin as more likely to have committed an immoral act (Grewal, 2017). The issue is further compounded by the fact that black boys may be easily misjudged, since they are commonly perceived as older and less innocent than their white counterparts (Goff et al., 2014), and young black men are seen as more physically threatening than their physical characteristics warrant (Wilson et al., 2017). All these biases in turn, become manifested in the unjust treatment of certain groups of people. For instance, in the United States there is documentation that there are significant disparities driven by racial stereotypes held by teachers and that these drive negative responses to Black students in the form of excessive and inequitable school discipline (e.g. Fenning & Rose, 2007; Gregory & Weinstein, 2008; Lewis et al., 2010). These in turn contribute to school failure and can also play a role in future incarcerations (Okonofua & Eberhardt, 2015; Rocque & Paternoster, 2011). Differential approaches by many teachers to their non-white students are, incredibly, manifested as early as in preschool (Hayes-Greene & Love, 2018). Naturally, with time the distinctions are magnified, and the stakes can become the literal difference between life and death.

Further exacerbating these disparities, overwhelmingly white school districts received $23 billion more than predominantly non-white school districts in state and local funding despite serving roughly the same number of children (EdBuild, 2019). If these are not *de jure* inequities, they are certainly systemic in origin. In the United States, about 7,600 schools have more than 75% white students, and about 1,200 have more than 75% of students who were not white. The non-white districts, tended to be much larger facilities than the white districts, but both served about the same number of students, 12.8 million children in non-white districts and 12.5 million in white districts. However, non-white districts took in about $54 billion in 2016 in local tax dollars, or about $4,500 per student. While white districts, home to higher incomes and less poverty, collected more than $77 billion, or just over $7,000 per student. The overall gap in state and local funding totaled $23 billion. Furthermore, teachers at schools with more Latinx and black children are paid on average $2,500 less than teachers in the district as a whole (U.S. Department of Education, 2011).

Such gaps in teacher pay, educational resources, physical plants, student agency and autonomy have very real consequences for racially minoritized students. As we have seen, non-white students are usually educated with other non-white students in overcrowded conditions, with limited resources, by less experienced teachers—in addition to the numerous other potential problems related to poverty, including nutrition, health care access, housing conditions, and increased stress. Yet, schools with non-white majorities are supposed to compensate for all the evils of poverty during this developmentally sensitive time period with fewer resources than are available to majority white schools. This is a lose-lose dynamic, and another facet of oppression of non-white communities.

**Mortality/Health**. Furceri et al. (2020) found that the five major pandemics of the past two decades, SARS (2003), H1N1 (2009), MERS (2012), Ebola (2014), and Zika (2016), all raised income inequality and ultimately raised inequalities for BIPOC. And now, as case counts rise for COVID-19 in the United Sates, the process is revealing exactly how much American inequality is shaped by racial differences. Long-standing systemic health and social inequities have resulted in the increased risk of getting COVID-19 or experiencing severe illness for racial and ethnic minority groups of all ages. At the height of the pandemic (July 2020), age-adjusted hospitalization rates for Covid-19 were highest among American Indian or Alaska Native, Black, and Latinx persons (CDC, 2020).

Social justice transcends economic equality and equity; it is a matter of life and death. It affects the way people live, their consequent chance of illness, and their risk of premature death, a clear demonstration of inequity. According to the World Health Organization (2008), a girl born today can expect to live for more than 80 years if she is born in Japan, Sweden, or the United States—but less than 57 years if she is born in Sierra Leone, Swaziland and Lesotho. Importantly, there are also dramatic differences in health that are closely linked with degrees of social disadvantage within the same country. For example, in the United States the current average life expectancy for a black person is 74.9 years, compared with 78.8 years for a white person (Arias & Xu, 2019). Similarly, black women are more than three times more likely to die from pregnancy-related causes than white women (Petersen et al., 2019), and infant mortality is highest for black women with a rate of 10.75 per 1,000 live births, while the rate for white infants is less than half at 4.63 (Ely & Driscoll, 2020).

Preventable health inequalities arise because of the circumstances in which people grow, live, work, and age, and the systems put in place to deal with healthcare. For example, according to the CDC, 13.8% of Blacks (Centers for Disease Control and Prevention, 2017a) and 10% of Latinx (Centers for Disease Control and Prevention, 2017b) reported having fair or poor health compared with 8.3% of non-Hispanic whites. Similarly, Latinx women are 40% more likely to have cervical cancer and 20% more likely to die from cervical cancer than non-Hispanic white women (HHS [U.S. Department of Health and Human Services], 2020), and Blacks have the highest mortality rate for all cancers combined compared with any other racial and ethnic group (Centers for Disease Control and Prevention, 2017b).

The conditions in which people live and die are shaped by political, social, and economic forces. Social and economic policies have a defining impact on whether a child can grow and develop full potential and live a flourishing life, or if its life will be blighted (Tulchinsky & Varavikova, 2015). The latest reported figures also show that in the United States, approximately 4.3 million children do not have health insurance (U.S. Census Bureau, 2019b) and this indicator impacts Latinx children at a higher rate of 8.2% uninsured, and American Indian/Alaskan Native of 13.2% uninsured, which are significantly higher than the 5.2% of white children without health insurance (Alker & Roygardner, 2019).

## Conclusion

In northern RI, there are two elementary schools separated by a mile, but they could be in different countries. The elementary school in Central Falls (n=417 students), is 16% white, 95% free/reduced lunch (National Center for Education Statistics, 2020a). This contrasts with the elementary school in nearby Lincoln, (n=272 students) which is 80% white, and 24% free/reduced lunch (National Center for Education Statistics, 2020b). According to the U.S. Census Bureau (2020b), the median value of owner-occupied housing units in Central Falls is $147,800, with just 19.4% of those homes being owner-occupied. The median household income in Central Falls is $31,724. Meanwhile in Lincoln, wealth indicators more than double as it has a median value of owner-occupied housing units of $303,700 with 67.3% of those homes being owner-occupied, and a median household income of $74,123. The stark differences between these two adjacent cities cannot be explained by any physical separation as no such divider exists. The differences must be related to other factors—a lack of social justice that ultimately forced a school to protest its own blighted circumstances.

*As I stood side-by-side with the other protesters, there was a sense of significance in our plight. This was a collective effort where teachers, students, parents, administrators, and local elected officials came together for the cause of social justice. However, nothing was decided that day, nor the following months, and no additional funding was secured. Two months after the rally, it became clear that those with political and economic power retained the ability to control the lives of those with neither.*

*Even with its meritocratic aspirations, American society is stratified, based on income, wealth, power, culture, behavior, and, perhaps most significantly, race/ethnicity. There are large racially based variations in wealth, material possessions, authority, access to education, access to healthcare, and treatment by the law-enforcement/judicial systems. Beginning with a past where people were valued unequally, persistent privilege and oppression based on race or ethnicity (Bernal, 2002; Giroux, 1982) continue to dominate every aspect of United States life. If we really do adhere to the iconic language of our founding documents, we must eradicate the inequalities that have created disparities in wealth and power. And here, then, is the core of this book: life science education focused on evolutionary biology, has a key contribution to make to social equality. Biology education centered around evolution provides a conduit to investigate social injustices and biological truths. Teachers can and must serve as agents for change and counteract the injustices of the past by using their institutional and academic roles to create systemic empowerment.*

## Notes

1 Most notably, the French Revolution, whose famous call to arms, "Liberté, égalité, fraternité," reflected Rousseau's formulation of popular equality and sovereignty.

2 Particularly seen with the development and influence of the post-Marxian critiques of dominant social paradigms and power structures pioneered by critical theorists and activists in the post-WWII globalized environment (see Bales, 2018).
3 A religious argument has also been made to support this kind of racism; we address this in Chapter 8.
4 We note, however, that the voting restrictions currently being implemented in many states in the United States, the result of the discredited claim that the 2020 Presidential election was "stolen," might be recognizable as examples of *de facto* and *de jure* oppression.
5 At least 70 people have died in police custody over the past decade after saying, "I can't breathe," and over half of them were Black (Baker et al., 2020).

## References Cited

Alker, J., & Roygardner, L. (2019). *The Number of Uninsured Children is on the Rise*. Georgetown University Center for Children and Families. Retrieved July 2, 2020, from https://ccf.georgetown.edu/wp-content/uploads/2019/10/Uninsured-Kids-Report.pdf

Arias, E., & Xu, J.Q. (2019). *United States Life Tables, 2017*. National Vital Statistics Reports. Hyattsville, MD: National Center for Health Statistics. www.cdc.gov/nchs/data/nvsr/nvsr68/nvsr68_07-508.pdf

Baker, M., Valentino-DeVries, J., Fernandez, M., & LaForgia, M. (2020, June 29). Three Words. 70 Cases. The Tragic History of "I Can't Breathe." *The New York Times*. www.nytimes.com/interactive/2020/06/28/us/i-cant-breathe-police-arrest.html

Baker, P. (2016, November 27). *The Tragic, Forgotten History of Black Military Veterans*. Retrieved July 29, 2020, from www.newyorker.com/news/news-desk/the-tragic-forgotten-history-of-black-military-veterans

Bales, S. (2018). *Social Justice and Library Work. A Guide to Theory and Practice*. Cambridge, MA: Elsevier.

Bernal, D.D. (2002). Critical Race Theory, Latino Critical Theory, and Critical Race-Gendered Epistemologies; Recognizing Students of Color as Holders and Creators of Knowledge. *Qualitative Inquiry*, *9*(1), 105–126.

Blakemore, E. (2019, June 21). *How the GI Bill's Promise Was Denied to a Million Black WWII Veterans*. Retrieved July 29, 2020, from www.history.com/news/gi-bill-black-wwii-veterans-benefits

Bronson, J., & Carson, E.A. (2019). *Prisoners in 2017*. Washington, DC: Bureau of Justice Statistics. www.bjs.gov/index.cfm?ty=pbdetail&iid=6546

CDC. (2020, June 25). *COVID-19 in Racial and Ethnic Minority Groups*. Retrieved July 30, 2020, from www.cdc.gov/coronavirus/2019-ncov/need-extra-precautions/racial-ethnic-minorities.html

Centers for Disease Control and Prevention. (2017a). *Health of Black or African American Non-Hispanic Population*. www.cdc.gov/nchs/fastats/black-health.htm

Centers for Disease Control and Prevention. (2017b). *Health of Hispanic or Latino Population*. www.cdc.gov/nchs/fastats/hispanic-health.htm

Civil Rights Data Collection: Central Falls. (2020). Retrieved July 13, 2020, from https://ocrdata.ed.gov/Page?t=d&eid=27134&syk=8&pid=2278

Clark, J. (2014). Addressing the Achievement Gap in the United States. In J. Clark (Ed.), *Closing the Achievement Gap from an International Perspective*. Dordrecht: Springer. https://doi.org/10.1007/978-94-007-4357-1_4

Cramme, O., & Diamond, P. (2009). Rethinking Social Justice in the Global Age. In O. Cramme & P. Diamond (Eds.), *Social Justice in the Global Age*, pp. 3–20. Malden, MA: Polity Press.

Cullinane, A. (2020, June 17). *Central Falls Teachers, Students, and Advocates Rally for More Funding*. Retrieved July 2, 2020, from https://turnto10.com/i-team/schools-in-crisis/central-falls-teachers-students-and-advocates-rally-for-more-funding

Damrosch, L. (2012, July–August). *Friends of Rousseau*. Retrieved August 3, 2020, from www.neh.gov/humanities/2012/julyaugust/feature/friends-rousseau

Darity, W.A., & Mullen, A.K. (2020). *From Here to Equality: Reparations for Black Americans in the Twenty-First Century*. Chapel Hill: The University of North Carolina Press.

Darling-Hammond, L. (2014). Closing the Achievement Gap: A Systemic View. In J. Clark (Ed.), *Closing the Achievement Gap from an International Perspective*. Dordrecht: Springer. https://doi.org/10.1007/978-94-007-4357-1_2

Darling-Hammond, L. (2016, July 28). *Unequal Opportunity: Race and Education*. Retrieved August 13, 2020, from www.brookings.edu/articles/unequal-opportunity-race-and-education/

Davis, D.M. (2020, July 21). One of the Only 4 Black Fortune 500 CEOs Just Stepped Down—Here Are the 3 that Remain. *Business Insider*. www.businessinsider.com/there-are-four-black-fortune-500-ceos-here-they-are-2020–2

EdBuild. (2019). *$23 Billion*. Jersey City, NJ: Author. https://edbuild.org/content/23-billion/full-report.pdf

Edwards, F., Lee, H., & Esposito, M. (2019). Risk of Being Killed by Police Use of Force in the United States by Age, Race–Ethnicity, and Sex. *Proceedings of the National Academy of Sciences of the United States of America*, *116*(34), 16793–16798.

Ely, D.M., & Driscoll, A.K. (2020). Infant Mortality in the United States, 2018: Data from the Period Linked Birth/Infant Death File. *National Vital Statistics Reports, 69*(7). Hyattsville, MD: National Center for Health Statistics.

Equal Justice Initiative. (2017). *Targeting Black Veterans: Lynching in America: EJI Report*. Retrieved July 29, 2020, from https://eji.org/reports/targeting-black-veterans/

Federal Housing Authority. (1936, April 1). *Underwriting and Valuation Procedure Under Title II of the National Housing Act with Revisions*, Part II, Section 2. Washington, DC. https://babel.hathitrust.org/cgi/pt?id=mdp.39015018409246&view=1up&seq=211&q1=incompatible%20racial

Fenning, P., & Rose, J. (2007). Overrepresentation of African American Students in Exclusionary Discipline: The Role of School Policy. *Urban Education, 42*, 536–559.

FRED. Federal Reserve Bank of St. Louis. (2019). *Distributional Financial Accounts*. Retrieved July 2, 2020, from https://fred.stlouisfed.org/release?rid=453

Furceri, D., Loungani, P., Ostry, J.D., & Pizzuto, P. (2020, May 8). *COVID-19 Will Raise Inequality if Past Pandemics Are a Guide*. Retrieved July 2, 2020, from https://voxeu.org/article/covid-19-will-raise-inequality-if-past-pandemics-are-guide

Gillborn, D., Demack, S., Rollock, N., & Warmington, P. (2017). Moving the Goalposts: Education Policy and 25 Years of the Black/White Achievement Gap. *British Educational Research Journal, 43*, 848–874.

Giroux, H.A. (1982). *Theory and Resistance in Education: A Pedagogy for the Opposition*. Boston: Bergin & Garvey.

Goff, P.A., Jackson, M.C., Di Leone, B.A.L., Culotta, C.M., & DiTomasso, N.A. (2014). The Essence of Innocence: Consequences of Dehumanizing Black Children. *Journal of Personality and Social Psychology, 106*(4), 526–545.

Gregory, A., & Weinstein, R. (2008). The Discipline Gap and African Americans: Defiance or Cooperation in the High School Classroom. *Journal of School Psychology, 46*, 455–475.

Grewal, D. (2017, January 17). *The "Bad Is Black" Effect*. Retrieved July 2, 2020, from www.scientificamerican.com/article/the-bad-is-black-effect/

Gruver, J. (2019, May 7). *Racial Wage Gap for Men—Compensation Research*. Retrieved August 4, 2020, from www.payscale.com/data/racial-wage-gap-for-men

Hamilton, A. (1787/2008). *The Federalist Papers: No. 7*. Retrieved August 7, 2020, from https://avalon.law.yale.edu/18th_century/fed07.asp

Hayes-Greene, D., & Love, B.P. (2018). *The Groundwater Approach: Building a Practical Understanding of Structural Racism*. Greensboro, NC: The Racial Equity Institute.

Herrnstein, R.J., & Murray, C. (1994). *The Bell Curve: Intelligence and Class Structure in American Life*. New York: Simon and Schuster.

HHS [U.S. Department of Health and Human Services]. (2020). *Cancer and Hispanic Americans*. Retrieved July 28, 2020, from https://minorityhealth.hhs.gov/omh/browse.aspx?lvl=4&lvlid=61

Hume, D. (1740/2000). *A Treatise of Human Nature*. D.F. Norton & M.J. Norton (Eds.). Oxford: Oxford University Press.

International Monetary Fund. (2020). *World Economic Outlook Database April 2020*. Retrieved July 2, 2020, from www.imf.org/external/pubs/ft/weo/2020/01/weodata/weorept.aspx?sy=2014

Jargowsky, P. (2015). *Architecture of Segregation: Civil Unrest, the Concentration of Poverty, and Public Policy*. The Century Foundation. Retrieved July 2, 2020, from https://production-tcf.imgix.net/app/uploads/2015/08/07182514/Jargowsky_ArchitectureofSegregation-11.pdf

Jencks, C., & Phillips, M. (2016, July 28). *The Black-White Test Score Gap: Why It Persists and What Can Be Done*. Retrieved August 13, 2020, from www.brookings.edu/articles/the-black-white-test-score-gap-why-it-persists-and-what-can-be-done/

Kiel, P., & Fresques, H. (2017, September 27). *Data Analysis: Bankruptcy and Race in America*. Retrieved July 29, 2020, from https://projects.propublica.org/graphics/bankruptcy-data-analysis

Ladson-Billings, G. (2006). From the Achievement Gap to the Education Debt: Understanding Achievement in U.S. Schools. *Educational Researcher, 35*(7), 3–12.

Lewis, C.W., Butler, B.R., Bonner, F.A., & Joubert, M. (2010). African American Male Discipline Patterns and School District Responses Resulting Impact on Academic Achievement: Implications for Urban Educators and Policy Makers. *Journal of African American Males in Education, 1*(1), 7–25.

Logan, J., & Stults, B. (2011). *The Persistence of Segregation in the Metropolis: New Findings from the 2010 Census*. Providence, RI: US2010 Project.

Lynn, R. (2006). *Race Differences in Intelligence: An Evolutionary Analysis*. Arlington, VA: Washington Summit Publishers.

McCarthy, G.E. (1990). *Marx and the Ancients: Classical Ethics, Social Justice, and Nineteenth Century Political Economy*. Lanham, MD: Rowman & Littlefield.

Mishel, L., & Wolfe, J. (2019, August). *CEO Compensation Has Grown 940% Since 1978: Economic Policy Institute*. Retrieved July 2, 2020, from www.epi.org/publication/ceo-compensation-2018/

National Center for Education Statistics. (2017). *Digest of Education Statistics: Children Living in Poverty*. Retrieved July 2, 2020, from https://nces.ed.gov/programs/raceindicators/indicator_rad.asp

National Center for Education Statistics. (2019). *Public High School Graduation Rates*. https://nces.ed.gov/programs/coe/pdf/Indicator_COI/coe_coi_2019_05.pdf; https://nces.ed.gov/programs/digest/d19/tables/dt19_104.10.asp?current=yes

National Center for Education Statistics. (2020a). *Ella Risk Elementary School*. Retrieved July 30, 2020, from https://nces.ed.gov/ccd/schoolsearch/school_detail.asp?Search=1&DistrictID=4400120&ID=440012000027

National Center for Education Statistics. (2020b). *Saylesville Elementary*. Retrieved July 30, 2020, from https://nces.ed.gov/ccd/schoolsearch/school_detail.asp?Search=1&DistrictID=4400570&ID=440057000139

Okonofua, J.A., & Eberhardt, J.L. (2015). Two Strikes: Race and the Disciplining of Young Students. *Psychological Science*, *26*(5), 617–624.

Orfield, G. (2008, Spring). Race and Schools: The Need for Action, Civil Rights Project/Proyecto Derechos Civiles, University of California–Los Angeles. *Research Brief from the NEA Research Visiting Scholars Series*, *1b*.

Paine, T. (1797). *Agrarian Justice: Social Security History*. www.ssa.gov/history/paine4.html

Petersen, E.E., Davis, N.L., Goodman, D., Cox, S., Syverson, C., Seed, K., Shapiro-Mendoza, C., Callaghan, W.M., & Barfield, W. (2019). Racial/Ethnic Disparities in Pregnancy-Related Deaths—United States, 2007–2016. *MMWR Morbidity and Mortality Weekly Report*, *68*, 762–765. https://doi.org/10.15585/mmwr.mm6835a3

Prilleltensky, I., & Gonick, L. (1996). Polities Change, Oppression Remains: On the Psychology and Politics of Oppression. *Political Psychology*, *17*(1), 127–148.

Rawls, J.A. (1999). *A Theory of Justice: Revised Edition*. Cambridge, MA: Belknap Press.

Reeves, R.V. (2017). *Dream Hoarders: How the American Upper Middle Class Is Leaving Everyone Else in the Dust, Why That Is a Problem, and What to Do about It*. Washington, DC: Brookings Institution Press.

Reeves, R.V., & Halikias, D. (2017, August 15). *Race Gaps in SAT Scores Highlight Inequality and Hinder Upward Mobility*. Retrieved August 13, 2020, from www.brookings.edu/research/race-gaps-in-sat-scores-highlight-inequality-and-hinder-upward-mobility/

Rehavi, S.B., & Starr, M.M. (2012). *Racial Disparity in Federal Criminal Charging and Its Sentencing Consequences*. University of Michigan Law & Economics, Empirical Legal Studies Center Paper No. 12–002. Available at SSRN: https://ssrn.com/abstract=1985377

Reiss, R. (2018, February 27). Latino CEOs Share Insights on Business Success. *Forbes*. www.forbes.com/sites/robertreiss/2018/02/27/latino-ceos-share-insights-on-business-success/#bf6126e2264e

Richwine, J. (2009). *IQ and Immigration Policy*. Doctoral dissertation, Harvard University. ProQuest Dissertations Publishing. Retrieved August 13, 2020, from https://delong.typepad.com/pdf-1.pdf

Robinson, C.J. (2000). *Black Marxism: The Making of the Black Radical Tradition*, 2nd ed. Chapel Hill, NC: The University of North Carolina Press.

Rocque, M., & Paternoster, R. (2011). Understanding the Antecedents of the "School-to-Jail" Link: The Relationship between Race and School Discipline. *The Journal of Criminal Law and Criminology*, *101*, 633–666.

Rosen, A.D. (1993). *Kant's Theory of Justice*. Ithaca, NY: Cornell University Press.

Ross, C.T. (2015, November 5). A Multi-Level Bayesian Analysis of Racial Bias in Police Shootings at the County-Level in the United States, 2011–2014. *PLoS One*. https://doi.org/10.1371/journal.pone.0141854

Rothstein, R. (2017). *The Color of Law. A Forgotten History of How Our Government Segregated America*. New York: Liveright Publishing Corporation.

Rousseau, J.-J. (1762/2004). *The Social Contract*. New York: Penguin Books.

Rusk, D. (1999). *Inside Game/Outside Game: Winning Strategies for Saving Urban America*, p. 90. Washington, DC: Brookings Institution Press.

Tax Policy Center. (2019, March 11). *Family Holdings of Wealth and Debt Can Vary Significantly by Race. The Median Net Worth of Non-Hispanic White Families Is 8 to 10 Times Greater Than for Both Non-Hispanic Blacks or African Americans and Hispanics or Latinos*. Retrieved August 2, 2020, from www.taxpolicycenter.org/fiscal-fact/median-value-wealth-race-ff03112019

Thompson, J. (2019). The GI Bill Should've Been Race Neutral, Politicos Made Sure It Wasn't. *Military Times*. www.militarytimes.com/military-honor/salute-veterans/2019/11/10/the-gi-bill-shouldve-been-race-neutral-politicos-made-sure-it-wasnt/

Tulchinsky, T.H., & Varavikova, E.A. (2015). *Expanding the Concept of Public Health in the New Public Health*, 3rd ed. San Diego: Elsevier Academic Press.

United Nations. (1948). *Universal Declaration of Human Rights*. Retrieved July 2, 2020, from www.un.org/en/universal-declaration-human-rights/

United Nations. (2006). *Social Justice in an Open World: The Role of the United Nations*. Retrieved July 2, 2020, from www.un.org/esa/socdev/documents/ifsd/SocialJustice.pdf

U.S. Census Bureau. (2019a). *Income and Poverty in the United States: 2018*. Retrieved July 2, 2020, from www.census.gov/content/dam/Census/library/publications/2019/demo/p60-266.pdf

U.S. Census Bureau. (2019b). *Health Insurance Coverage in the United States: 2018*. Retrieved July 2, 2020, from https://www2.census.gov/programs-surveys/demo/tables/p60/267/tableA1.pdf

U.S. Census Bureau. (2020a). *QuickFacts: United States, Latinos*. Retrieved July 2, 2020, from www.census.gov/quickfacts/fact/table/US/RHI725219

U.S. Census Bureau. (2020b). *Lincoln Town, Providence County, Rhode Island; Central Falls City, Rhode Island; United States*. Retrieved July 30, 2020, from www.census.gov/quickfacts/fact/table/lincolntownprovidencecounty rhodeisland,centralfallscityrhodeisland,US/PST045219

U.S. Department of Education. (2011, September 27). *Significant Pay Gap for Teachers in Schools Serving More Latino and African-American Students, According to New U.S. Department of Education Data*. Retrieved August 8, 2020, from www.ed.gov/news/press-releases/significant-pay-gap-teachers-schools-serving-more-latino-and-african-american-students-according-new-us-department-education-data

U.S. Department of Housing and Urban Development. (2020). *The Annual Homeless Assessment Report to Congress*. https://files.hudexchange.info/resources/documents/2019-AHAR-Part-1.pdf

Walmsley, R. (2019). *World Prison Brief*. London: Institute for Crime & Justice Policy Research. www.prisonstudies.org/world-prison-brief-data

Western, B., & Pettit, B. (2010). Incarceration and Social Inequality. *Daedalus, 139*(3), 8–19.

Wilson, J.P., Hugenberg, K., & Rule, N.O. (2017). Racial Bias in Judgments of Physical Size and Formidability: From Size to Threat. *Journal of Personality and Social Psychology, 113*(1), 59–80. https://doi.org/10.1037/pspi0000092

World Health Organization. (2008). *Commission on Social Determinants of Health: Closing the Gap in a Generation: Health Equity through Action on the Social Determinants of Health*. Geneva. http://whqlibdoc.who.int/publications/2008/9789241563703_eng.pdf

# 6

# Pedagogy Underlying the Integration of Racial Justice into High School Biology Curricula

*It was the middle of April 2011 and I had prepared my speech for those two students. We stood waiting for the results to be called and they had already called the third and second place winners. These two students had worked so much, they stayed after school so many days and even received helped from two medical students to train for the Anatomy & Physiology event of the Science Olympiad. I think the students still had hope, but I thought that third place was probably the best they could do. After all, this was the first year that the school had a team and in theory, these students from the most impoverished school in the state could not compete with all those private school students who had long been working towards this competition. The announcer made a dramatic pause before he screamed "CENTRAL FALLS HIGH SCHOOL"—the winners of the event. The whole place erupted. There was a standing ovation! My students and I hugged, and tears flowed from each of our faces. The impossible just happened—two immigrant students from the poorest school in the state had beaten every other competitor to become the gold medal winners.*

*I remember a teacher telling me when I first started the team, that I was wasting my time, "they don't even come for regular science class, you think they are going to come after school for more science?" But because of that win, I was able to leverage the district to allow me to teach Advanced Placement (AP) biology. The following academic year, 2011–2012, I started the AP biology class with 23 students. I thought that was too many for an AP class with an extensive laboratory requirement, so I decided to describe the seriousness of the commitment expected by students if they were going to succeed. I recall saying something like what my college biology professor, Lloyd Matsumoto, had said the first day of my introductory biology class, "if you are not willing to eat, sleep, think, and dream biology, then this is probably not*

DOI: 10.4324/9781003409144-9

*the class for you." I felt strange and uncomfortable saying it, but given the extremely limited materials and resources, I was aiming to only have the 15 most committed students stay. To my surprise, the next day five additional students joined the class. Instead of dropping out, when Central Falls students heard about this challenge, they invited others. That first year of AP biology we had to make do with what we had, and we needed to address deep gaps in knowledge in all of us. I do not think it is an understatement to say that I learned more than the students, including the realization that studying and understanding science was changing all of us in unexpected ways.*

*I pondered about bio class and how it challenged and rewarded all of us. What was it about that rigorous class that was empowering students? How could a class about science become a tool for my and my students' liberation?*

## Introduction

The Lessons prescribed in Chapters 2–4 are the result of merging published literature which addresses racial justice in the classroom, with work which documents ways in which some biology teachers have incorporated such ideas into their classes (Upegui et al., 2021). The intent is to provide a wide range of ideas which are designed to culminate in engaging students in critical thinking about these issues and ultimately in helping those students construct a more equitable society than the current one. Now more than ever, there is both support for (Collins, 2020), and backlash against,[1] teaching the history of US racism. We begin here with a brief review of the key pillars upon which our idea of integrating social justice into high-school biology curricula stands.

### The Importance of Science Education

It was in World War I that science and scientists first became recognized as vital to national power, but it was the 1957 launch of Sputnik 1, the first satellite to orbit the Earth, which was perceived globally as a shot across the bow of post-World War II American world hegemony (Cadbury, 2006), and was seen in the United States as both a potential threat to American national security as well as a blow to national pride. This was the first leg of the "Space Race" between the United States and USSR, which ultimately culminated in the Apollo Moon landings. The development of missiles and rockets happened in parallel with the capacity to deliver nuclear weapons, to spy on the enemy and therefore to wield power (Cadbury, 2006). In the United States, with the USSR having successfully penetrated space, it was determined that there was a shortage of scientists with too few qualified people entering the field (Wissehr et al., 2011).

At that time, therefore, science and science education became American priorities, and in 1958 Congress passed the National Defense Education Act (NDEA) which authorized increased funding for scientific research as well as education in science and mathematics. More recently, the impetus for scientific education has gained renewed support in the form of STEM (Science, Technology, Engineering and Math) education. STEM is currently at the forefront of many educational campaigns (Ad Council, 2018) and funding opportunities from the U.S. Department of Education (2018). Former President Barack Obama stated that "[Science] is more than a school subject, or the periodic table, or the properties of waves. It is an approach to the world, a critical way to understand and explore and engage with the world, and then have the capacity to change that world" (Obama, 2015). The sentiments are indicative of a larger trend in which education, and in particular science education, has come to symbolize a competitive and successful society. In our society science and the resulting technology from scientific endeavors enjoy widespread value. Historically, science has been used for many purposes (as we have seen previously in this book), including winning wars, but unsurprisingly, good science can also be instrumental in challenging racial inequities.

## Teaching Science

Given the importance of science education to the society,[2] it is imperative that we find ways to ensure productive and constructive methods to teach and learn science, because the methods by which we teach science are of critical importance individually and collectively. Sadler et al. (2016) found that students who were taught science that included socio-scientific issues (SSIs) demonstrated statistically and practically significant gains in content knowledge. SSIs provide students with a context to support their active examination of the relationships between science and their lives in society. Issues such as global climate change, genetic engineering, alternative energy, stem-cell research, and modern health care have direct impact on students' lives. A high level of science literacy is required to negotiate and make decisions regarding complex, social issues with theoretical and/or conceptual links to science (Fowler et al., 2009). This situated learning has been shown to foster content learning and improved performance on high-stakes tests (Sadler et al., 2016), and even help students develop moral sensitivity (Fowler et al., 2009).

Academic research advocates learning science in the context of SSIs for using science in daily lives. Ke et al. (2020) reported that students held positive attitudes towards SSI-based learning experiences and found SSI work to be relevant, interesting, promoting agency, and beneficial for their science, not to say their lives. Therefore, if one of the aims of science education is to increase students' active participation in society, particularly in the context of science-related social issues, then we must situate their learning within issues of relevance to them.

## Pedagogy

Pedagogy is arguably as important as curriculum. The teachers who provided much of the data that were used to inform this Chapter (see Upegui et al., 2021) offered insights about the methods they use in their classrooms. Inquiry- (or "problem-") based learning is one approach that parallels the methods and practices of professional scientists in order to construct knowledge (Keselman, 2003). In contrast with more traditional education approaches that rely on the teacher as the fact/knowledge keeper, inquiry-based learning places the teacher as facilitator rather than a lecturer (Pedaste et al., 2015). Inquiry-based teaching has the potential to help students develop and practice critical thinking and problem-solving skills (Dostál, 2015).

Formative work done by American educator and philosopher John Dewey and Russian psychologist and educator L. S. Vygotsky connected individual experiencing to the learning process that happens in schooling (since popularized as "experiential learning"). Independently, both authors describe experiences as more than mental constructs resulting from independent events had by individuals, but as transactions in and across space and time within irreducible person-in-setting units (Roth & Jornet, 2014). Experience, then, as treated by Dewey and Vygotsky, does not simply refer to the participation in events or activities that result in having certain feelings. Instead, for them, experience is a category of thinking, or a minimal unit of analysis that includes the intellectual, affective, and practical characteristics of the individual. Experiences entail people's material and social environment, and their transactional relations and affect (Roth & Jornet, 2014). According to Vygotskij (1935/2001), experiences are not something hidden away within individuals, but encompass the space and time across individuals and their societal relations, which themselves are perfused with affect. Similarly, Dewey (1934/2008) emphasized the continuous nature of experience and described the importance of student reflection in their learning experiences.

Building on Dewey and Vygotsky's work, Pugh et al. (2017) define transformative experiences as learning episodes in which students use ideas from the science classroom to see and experience the world differently in their everyday lives. Their construct is defined by three characteristics: motivated use, expansion of perception (seeing objects, events, or issues through the lens of the content), and experiential value (valuing content for how it enriches everyday experience). It is these types of transformative experiences that our students need the most to pursue just lives.

## Storytelling in Pedagogy

Perhaps the single most important pedagogical activity presented here is the process of storytelling. Storytelling is undoubtedly one of the oldest forms of communication where humans teach, influence, inspire, and engage others in learning. Singh and Brownell (2019) describe storytelling as "an art form through which people of all ages are able to connect thoughts and ideas to meaningful emotions" (p. 109). In the storytelling activity, students will tell their story and listen to each other's stories. The process of listening to stories provides opportunities to make connections, to expand their thinking and to collaborate with others.

Seasoned teachers utilize storytelling as a major technique to connect with their students and promote the exploration of racial justice in their classrooms (Upegui et al., 2021). Calabrese Barton (2003) wrote that

> [S]tories are powerful research tools. They provide us with a picture of real people in real situations, struggling with real problems. They banish the indifference often generated by samples, treatments, and faceless subjects. They invite us to speculate on what might be changed and with what effect.
>
> (p. 7)

In that vein, Immordino-Yang (2016) provides neurobiological evidence that compelling narratives can trigger emotions that may activate a mindset of "imagined possible futures" (p. 6) that can in turn shape one's identity. Scientific understanding of the influence of emotions on thinking and learning reveals that emotion and cognition are supported by interdependent neural processes (Immordino-Yang, 2016). This is particularly important for teachers who want to help students build memories, engage complex thoughts, and critically analyze problems. Put succinctly, effective teachers must find ways to engage students in issues that matter to them.

The teachers interviewed in Upegui et al. (2021) uniformly used storytelling to engage the students both emotionally and intellectually and help them make personally relevant meaning of the issues involving racial and social justice that they encountered in their daily lives. If we want to improve the lives of our students individually and *en masse*, our goals as educators must go beyond the mastery of content knowledge and logical reasoning skills in the traditional academic sense. Our students must be able to use these academic skills and knowledge outside of the structured context of a school (Immordino-Yang & Damasio, 2007).

## Why Evolutionary Biology?

The simplest answer to the question, "Why evolutionary biology?" for the introduction of racial justice in biology classes is that ultimately both evolutionary biology and race are about our understanding of diversity. The result of evolution is, of course, the generation of diversity, and the study of evolution ultimately leads to a refined understanding of relationships among organisms. The concept of "race" is also about the diversity within humanity; how it arose (that is, evolved); how we measure it; and finally, how closely (or distantly) humans are related to each other. Understanding the scientific significance of human diversity translates directly into the social fabric of relationships among humans.

Since the Enlightenment advent of science as an intellectual discipline, racists have always sought to justify their inequitable world views using science. However, as the quantity and quality of available science has grown, it has become more and more difficult to find reputable science that is even agnostic about supposed racial differences, and thus racists have had to go to extraordinary lengths to promote preconceived biological world views.

Uniformly high-quality science education is the only way to implement a "sophisticated scientific understanding in which everyone can share" (Lewontin, 1991, p. 16). Donovan (2017) argued quantitatively that a proper understanding of genetic variation has substantial implications for racial prejudice. Donovan et al. (2019) demonstrated that students who were taught biology with racist undertones became less interested in socializing across racial lines and "less supportive of policies that reduce racial inequality in education (p. 379)," than those who were not. Donovan et al. showed that students that have racist beliefs reinforced by biology curricula exhibited less support for educational policies that reduce racially based inequalities since in the minds of those students, such inequalities are fixed immutably by genetic content. Similarly, Willinsky (2020) found that current United States' biology textbooks do not adequately prepare students to discern how white supremacist interpretations of the meaning of "race" in science involve controversy, and inconsistency, and misinterpretation.

While there are real biological consequences to the lived experience of race, evolutionary biology states that the genetic variation *within* the groups we designate as races is significantly larger than the variation *between* those groups (Graves & Goodman, 2021). Genetic comparisons demonstrate that, in terms of DNA, human geomes are 99.9% similar. And even more dramatically, we now know that genes do not control behavior to the exclusion of other influences (e.g., if you have certain genotype, you must express those in the form of an immutable phenotypical trait); instead, there is a far more of complex relationship that includes environmental and malleable molecular triggers (Radick, 2020). But these complicated biological facts are commonly unknown or unappreciated, most people preferring instead more simplistic explanations, that support their preconceptions.

Evolutionary theory shows us that human genetic variation is real but is absolutely not the same as the commonly held idea of race (Graves & Goodman, 2021). Humans are genetically one of the most homogeneous species on the planet, an unsurprising fact when the human propensity for travel and trade are considered. We are more homogeneous than any other primates including chimpanzees, as described by Bowden et al. (2012): chimpanzees from different, but geographically-proximal populations, were substantially more different genetically than humans living on different continents. The majority of genetic difference between humans is within the individual level (Serre & Pääbo, 2004). In fact, humans are not biologically grouped in skin-color-coded discrete clusters, as has been suggested by the idea of race, but rather gradients in genetic variation that extend over the entire globe (Fuentes, 2021). Indeed, the greatest genetic diversity in human populations is to be found in sub-Saharan Africa.

## Teaching Evolutionary Theory

Because evolutionary theory shows us that human genetic variation is real but it is strikingly discordant with the commonly held idea of race (Graves & Goodman, 2021), it serves as a direct means of dismantling racist ideas. For teaching evolutionary theory, Pobiner et al. (2019) suggest that the pedagogical approaches with the most promise are the ones that utilize strategies aiming at conceptual change and attend to the relevance of understanding evolution. Pobiner et al. (2019) define three major categories of cognitive biases to learning about natural selection: essentialism,[3] intentionality,[4] and teleology.[5] These authors report evidence for the efficacy of a constructivist, guided-inquiry pedagogical approach using human evolution case studies to teach high school biology students about natural selection; this approach was best suited to overcome common cognitive biases and misconceptions. The constructivist-learning approach, which elicits student ideas and then gives multiple opportunities to build scientific explanations consistent with what is known about the subject, resulted in student gains in understanding evolution. Additional data also showed that at least in some contexts, explicitly drawing on students' cultural concerns through a discussion of non-scientific views can help students to gain a more correct and focused understanding of evolutionary science (Bertka et al., 2019). Moreover, Pobiner et al. (2018) showed that using examples of evolution in humans, created a more relevant and interesting experience to students, by comparison to examples of evolution in more distantly related organisms (which the students may have never encountered).

In a multiple case-study project that sought to explore ways in which biology teachers incorporate social justice into an already-packed curriculum that is dictated by the College Board, several key themes arose (Upegui et al., 2021). The data from that study suggest there are multiple ways to incorporate social justice in the AP biology classroom. For example, storytelling emerged as a consistent and powerful tool, particularly with types of stories that allow students to visualize either injustices or unscientific conclusions reached by racists. Additionally, the targeted and explicit teaching of the process of science (emphasizing the primacy of data over assertion), as well as a pedagogical connection between social justice and science in the biology curriculum, enable students to use science as a way to challenge limiting social constructs and build social justice in their lives. Herein lies the power of targeted narratives, since we know that disassociated facts do not generally change the minds of people. Therefore, we need stories that explain science as a process and subsequently enable students to explore social injustices. Benefits from this approach include the recognition of science as an essential analytical tool to participate in the creation of a more equitable society, rather than a suite of experimental procedures and results to be memorized and regurgitated.

## Assessment

Lastly, not everything that matters can be measured, but since our current education system often demands the quantification of score change as evidence for learning, some teachers may feel inclined to directly assess the impact of the activities in this book in student learning. There are several options for additional measurements aside from the numerous "check for understanding" assessments that are provided within the module. For example, teachers may be apt to use a pre/post assessment like the "Implicit Association Test" from Harvard University (https://implicit.harvard.edu/implicit/) to evaluate changes in students' implicit scores as indication of thoughts and feelings outside of conscious awareness. Similarly, teachers can use the validated instrument developed by Donovan et al. (2020) to assess changes in genetic essentialist beliefs which are associated with racists ideologies. Still, the most significant and meaningful result of this module will be longer-term, felt by the students and their community.

*Coming out of that first AP bio class, in April of 2012, David Hernandez and Bintou Camara competed in the Anatomy event of the Science Olympiad. They came in second place and while waiting for the results, Bintou said something that stuck out to me: "Now they know that we are for REAL. We ain't playing!" She was right, playing is not what we were doing, we were doing science. But this science was empowering students in ways I had not predicted. David was the first student in the history of our school that was accepted into the Program in Liberal Medical Education at Brown University. David finished medical school and Bintou is a labor and delivery nurse.*

*Edwin Africano was also in that first AP bio class, but he was not interested in further study of anatomy. That year he scored a 5 on the AP exam and told me that he was interested in engineering. He became the first student in our school to graduate from the Massachusetts Institute of Technology. The following year, Jean Paul Valencia came to me after class and discussed his plans. He said, "We already have Brown and MIT, I want to go to Harvard!" I noticed some hesitation in his voice, so I asked him about it. He confided, "I don't know if I can get into Harvard." So, I asked him the regular question posed in our science classroom, "What evidence do you have to support that statement?" He looked around for an answer that would satisfy me, and came up with, "Our school is over 100 years old, and no one from our school has ever gotten into Harvard." I followed up, "But we know that lack of evidence is not the same thing as evidence that negates or supports a hypothesis."*

*On his own, Jean Paul offered this connection, "Mr. Upegui you know how in class you told us that genes are written in stone, but their reading is flexible, like plastic? And you told us that tomato plants that have the same genes can have different phenotypes if the conditions are different? So that even if the plants are genetically the same, but you change access to sunlight, water, soil nutrients and air quality, then you can't expect the same quality or quantity of fruit, right? So, I think that the students here have similar genetic make-up and potential to other students from wealthier communities, but the conditions have turned against us." There it was, we weren't playing with this science thing—but instead were empowering students. I had not heard it out loud before Jean Paul said it to me. Jean Paul graduated from Harvard in 2017.*

## Notes

1 During production of this book, these issues took on new significance with Florida governor Ron DeSantis' revision of the state's African-American Studies AP course—both in state secondary and tertiary educational institutions—because "the content of this course is inexplicably contrary to Florida law and significantly lacks educational value." The law to which the content of the course was "contrary" was the previous year's "Stop wrongs to our Kids and Employees Act" (informally dubbed the

"Stop WOKE Act") whose intent was to take "a stand against the state-sanctioned racism that is critical race theory" (quoted in The Hill TV, February 07, 2023). But as has been so-often noted; the objection is not really to Critical Race Theory (CRT), a subject which is inappropriate for, and rarely—if ever—introduced at, the high school or early college level. The resistance comes from a feeling that to teach a more racially balanced view of American social history somehow threatens and undermines the social fabric of the United States because it makes some groups feel badly about their history. The racially balanced perspective has been misleadingly dubbed CRT. While it is true that the roots of CRT are to be found in a racially balanced understanding of American history and in an acknowledgement the systemic racism that we have documented in Chapters 5–10 in this book, CRT itself is something very different (see Crenshaw et al., 1993)

2 A point that is particularly topical since, as this is being written, the United States is three years and 102 million cases with 1.12 million deaths, into a pandemic, and for all the life-saving potential of vaccines, there is strong, widespread distrust in, and resistance to, their use. Lawn signs now affirm the importance of science, a viewpoint that would have been self-evident 20 years ago.

3 *Essentialism* is the belief that individuals and groups have an essential nature that allows them to be placed into categories with sharp, immutable boundaries.

4 *Intentionality* assumes that events are caused by an intentional agent and are purposeful, goal-directed, or progressive, including the specious idea that evolution is progressing toward an ideal.

5 *Teleology* assumes that the characteristics organisms are explained by their ultimate purpose.

## References Cited

Ad Council. (2018). *Empowering Girls in STEM*. Retrieved March 13, 2021, from www.adcouncil.org/Our-Campaigns/Education/Empowering-Girls-in-STEM

Bertka, C., Pobiner, B., Beardsley, P., & Watson, W. (2019). Acknowledging Students Concerns Above Evolution: A Proactive Teaching Strategy. *Evolution: Education and Outreach, 12*.

Bowden, R., MacFie, T.S., Myers, S., Hellenthal, G., Nerrienet, E., Bontrop, R.E., Freeman, C., Donnelly, P., & Mundy, N.I. (2012). Genomic Tools for Evolution and Conservation in the Chimpanzee: Pan troglodytes ellioti is a Genetically Distinct Population. *PLoS Genetics, 8*(3), e1002504. https://doi.org/10.1371/journal.pgen.1002504

Cadbury, D. (2006). *Space Race: The Epic Battle Between America and the Soviet Union for Dominion of Space*. New York: HarperCollins.

Calabrese Barton, A. (2003). *Teaching Science for Social Justice*. New York: Teachers College Press.

Collins, J. (2020, October 13). *Poll Shows Support for Teaching History of U.S. Racism—With or Without Parental Consent*. Retrieved November 8, 2020, from www.the74million.org/article/poll-shows-support-for-teaching-history-of-u-s-racism-with-or-without-parental-consent/

Crenshaw, K., Gotanda, N., Peller, G., & Thomas, K. (1993). *Critical Race Theory: The Key Writings that Formed the Movement*, p. 494. New York: The New Press.

Dewey, J. (1934/2008). *The Later Works Volume 10: Art as Experience*. J.-A. Boydston (Ed.). Carbondale: Southern Illinois University Press. (Original work published in 1934).

Donovan, B.M. (2017). Learned Inequality: Racial Labels in the Biology Curriculum Can Affect the Development of Racial Prejudice. *Journal of Research in Science Teaching, 54*(3), 379–411.

Donovan, B.M., Semmens, R., Keck, P., Brimhall, E., Busch, K.C., Weindling, M., & Salazar, B. (2019). Towards a More Humane Genetics Education: Learning about the Social and Quantitative Complexities of Human Genetic Variation Research could Reduce Racial Bias in Adolescent and Adult Populations. *Science Education*, 1–32.

Donovan, B.M., Weindling, M., Salazar, B., Ducan, A., Stuhlsatz, M., & Keck, P. (2020). Genomics Literacy Matters: Supporting the Development of Genomics Literacy Through Genetics Education could Reduce the Prevalence of Genetic Essentialism. *Journal of Research in Science Teaching, 2020*, 1–31.

Dostál, J. (2015). *Inquiry-Based Instruction: Concept, Essence, Importance and Contribution*. Olomouc: Palacký University. https://doi.org/10.5507/pdf.15.24445076

Fowler, S.R., Zeidler, D.L., & Sadler, T.D. (2009). Moral Sensitivity in the Context of Socioscientific Issues in High School Science Students. *International Journal of Science Education, 31*(2).

Fuentes, A. (2021). "On the Races of Man": Race, Racism, Science, and Hope. In J. Desilva & J. Browne (Eds.), *A Most Interesting Problem: What Darwin's Descent of Man Got Right and Wrong About Human Evolution*, pp. 144–161. Princeton, NJ: Princeton University Press.

Graves, J.L., & Goodman, A. (2021). *Racism, Not Race: Everything You Wanted to Know About Race: But Were Afraid to Ask*. New York: Columbia University Press.

The Hill TV. (2023, February 7). *DeSantis Sparks Outrage with Rejection of African American Studies Class*. Retrieved February 07, 2023, from https://thehill.com/homenews/campaign/3821500-desantis-sparks-outrage-with-rejection-of-african-american-studies-class/

Immordino-Yang, M.H. (2016). *Emotions, Learning, and the Brain: Exploring the Educational Implications of Affective Neuroscience*. New York: W. W. Norton & Company.

Immordino-Yang, M.H., & Damasio, A.R. (2007). We Feel, Therefore We Learn: The Relevance of Affect and Social Neuroscience to Education. *Mind, Brain, and Education, 1*, 3–10. https://doi.org/10.1111/j.1751-228X.2007.00004.x

Ke, L., Sadler, T.D., Zangori, L., & Friedrichsen, P.J. (2020, May 23). Students' Perceptions of Socio-Scientific Issue-Based Learning and Their Appropriation of Epistemic Tools for Systems Thinking. *International Journal of Science Education, 42*(8), 1339–1361. https://doi.org/10.1080/09500693.2020.1759843

Keselman, A. (2003). Supporting Inquiry Learning by Promoting Normative Understanding of Multivariable Causality. *Journal of Research in Science Teaching, 40*, 898–921. https://doi.org/10.1002/tea.10115

Lewontin, R.C. (1991). *Biology as Ideology*. New York: HarperCollins.

Marks, J. (2010). Ten Facts about Human Variation. In M.P. Muehlenbein (Ed.), *Human Evolutionary Biology*, pp. 265–276. Cambridge: Cambridge University Press.

Obama, B. (2015). *Remarks by the President at White House Science Fair*. Retrieved March 15, 2021, from https://obamawhitehouse.archives.gov/the-press-office/2015/03/23/remarks-president-white-house-science-fair#:~:text=It's%20more%20than%20a%20school,to%20share%20this%20accumulated%20knowledge

Pedaste, M., Mäeots, M., Siiman, L.A., De Jong, T., Van Riesen, S.A., Kamp, E.T., & Tsourlidaki, E. (2015). Phases of Inquiry-Based Learning: Definitions and the Inquiry Cycle. *Educational Research Review, 14*, 47–61.

Pobiner, B., Beardsley, P.M., Berkta, C.M., & Watson, W.A. (2018). Using Human Case Studies to Teach Evolution in High School A.P. Biology Classrooms. *Evolution: Education and Outreach, 11*(3). https://doi.org/10.1186/s12052-018-0077-7

Pobiner, B., Watson, W.A., Beardsley, P.M., & Bertka, C.M. (2019). Using Human Examples to Teach Evolution to High School Students: Increasing Understanding and Decreasing Cognitive Biases and Misconceptions. In U. Harms & M.J. Reiss (Eds.), *Evolution Education Re-Considered: Understanding What Works*, pp. 185–205. Cham: Switzerland: Springer Nature.

Pugh, K.J., Bergstrom, C.M., & Spencer, B. (2017). Profiles of Transformative Engagement: Identification, Description, and Relation to Learning and

Instruction. *Science Education, 101*, 369–398. https://doi-org.uri.idm.oclc.org/10.1002/sce.21270

Radick, G. (2020). Making Sense of Mendelian Genes. *Interdisciplinary Science Reviews, 45*(3), 299–314. https://doi.org/10.1080/03080188.2020.1794387

Roth, W.M., & Jornet, A. (2014). Toward a Theory of Experience. *Science Education, 98*, 106–126. https://doi-org.uri.idm.oclc.org/10.1002/sce.21085

Sadler, T.D., William, L., Romine, W.L., & Topçu, M.S. (2016, July 2). Learning Science Content through Socio-Scientific Issues-Based Instruction: A Multi-Level Assessment Study. *International Journal of Science Education, 38*(10), 1622–1635. https://doi.org/10.1080/09500693.2016.1204481

Serre, D., & Pääbo, S. (2004). Evidence for Gradients of Human Genetic Diversity Within and Among Continents. *Genome Research, 14*(9), 1679–1685.

Singh, S., & Brownell, C. (2019). *Math Recess: Playful Learning in an Age of Disruption*. Canterbury: IMPress.

Upegui, D., Coiro, J., Battle, S., Kraus, R., & Fastovsky, D. (2021). Integration of the Topic of Social Justice into High School Biology Curricula. *Science & Education*. https://doi.org/10.1007/s11191-021-00287-yI

U.S. Department of Education Announces STEM, Computer Science Education Grant Opportunities. (2018, April 26). Retrieved November 23, 2018, from www.ed.gov/news/press-releases/us-department-education-announces-stem-computer-science-education-grant-opportunities

Vygotskij, L.S. (1935/2001). *Lekcii po pedologii [Lectures on Pedology]*. Izhevsk, Russia: Udmurdskij University (A translation of Lecture 5 to which we refer here exists as Vygotsky, L.S. (1994). The Problem of the Environment. In R. Veer & J. Valsiner (Eds.), *The Vygotsky Reader*, pp. 338–354. Oxford, England: Basil Blackwell). (Original work published in 1935).

Willinsky, J. (2020). The Confounding of Race in High School Biology Textbooks, 2014–2019. *Science & Education*. https://doi-org.uri.idm.oclc.org/10.1007/s11191-020-00104-y

Wissehr, C., Concannon, J., & Barrow, L.H. (2011). Looking Back at the Sputnik Era and Its Impact on Science Education. *School Science and Mathematics, 111*(7), 368–375.

# 7

# The Theory of Evolution by Natural Selection

*I spent a large part of my early childhood with Franciscan Missionaries on the Atlantic coast of Colombia because my aunts are Franciscan nuns and my immediate family had extremely limited economic resources. The missions were located throughout the northern part of the country and in remote places. This meant that I was able to wander and wonder about the natural fauna and flora around the Mission buildings. Some of my earliest memories include the "thrill" of swimming in ponds where caimans' eyes could be seen on the waterline staring back at us and gathering specimens of flowers and insects I would show my aunts.*

*Undoubtedly, I was captivated with the natural world around me, and because my aunts were also teachers, I was encouraged to ask questions. Alba, the older of the sisters, told me that even the smallest flowers display the marvelous intricacies of life. And Celly, the younger one, taught me that we are connected to every living creature. These were some of my early introductions to a version of evolutionary theory.*

*The lessons those nuns shared with me helped me organize my thinking about the natural world. They did not see science as being in opposition to their faith and instead, taught me to value the natural world and appreciate its beauty. For me, the concept of natural selection inspired a sense of wonder because it provided an explanation as to how the complexity and diversity of life on Earth can arise through a simple process. I found it beautiful and mysterious to comprehend how this simple idea can give rise to such complex phenomena. In this chapter we will provide an overview of the theory of evolution including some of the history of theoretical and scientific development that served as impetus to our current understanding of biological evolution.*

*Based on my observations, I, like many other people, thought about our place in the world and began to create working ideas that combined spirituality, hypotheses,*

DOI: 10.4324/9781003409144-10

*and experiments. At that time, these ideas were the immature stuff of childhood, and it was many years later that I finally would begin to formally learn about the use of science and the theory of evolution. But when we say evolution, what do we actually mean? Should we "believe" in the theory of evolution? What does the theory of evolution offer us?*

Of all the established scientific theories, perhaps none remains as controversial as the theory of evolution by natural selection. One would be much harder pressed to find widespread contentious objections to germ theory, cell theory, plate tectonics, quantum theory, special relativity, valence bond theory, and so many other scientific theories that provide insights into the natural world. But evolution, well, that is a different story; after more than a century and a half of validation, there is still passionate resistance to the theory, particularly as it has been interpreted to conflict with certain religious beliefs (Kitzmiller v. Dover, 2005). Yet, the consequences of organic evolution make it the central, single unifying concept in biology (Dobzhansky, 1973). Without the theory of evolution, biology would be an exercise in collecting disjointed facts.

Although aggressively contested, no modern scientific theory is misunderstood more than evolutionary theory (Thagard & Findlay, 2009). To explore why this is, we first describe why science is a reliable tool for exploring the natural world, and then revisit the outlines of evolutionary theory as it is understood by modern evolutionary biologists. Finally, we explain how, like any robust scientific theory, it has withstood significant scientific and philosophical testing.

## Science as a Means of Understanding the Natural World

The natural world is complex and to make sense of this complexity, humans have developed *science*, a word that is used to describe both the method of discovering reliable knowledge about the observable universe, and the body of knowledge gained through such inquiry. Astrophysicist Carl Sagan (1934–1996) described science as not just a body of knowledge but as a way of thinking. He viewed scientific thinking as both imaginative and disciplined, allowing us to reach towards an objective understanding of how the natural world is, rather than how we might wish to perceive it.[1]

Modern science was born during the Enlightenment, the western European intellectual movement of the 16th through 18th centuries, that emphasized logic and observation over mysticism and tradition. Resulting from Enlightenment thinking, natural phenomena are today understood by the establishment of scientific hypotheses, which make predictions that can be tested by observations. Hence, scientific propositions (hypotheses) must be falsifiable (Popper, 1994) and are intentionally understood to be ephemeral;

subject to modification or rejection as new data are adduced. Sagan (1997) distinguished science as working better than any other system of explanations, because it has a "built-in error-correcting machine" (p. 31). Science as characterized by philosopher of science Karl Popper (1902–1994) is a hypothesis-driven deductive system where nothing is ever "proven" but rather deduced hypotheses are either falsified or fail to be falsified. And because we cannot know every case that represents the phenomenon we are studying, additional cases that fail to falsify our ideas, do not further confirm our ideas, that is, inductively increase the likelihood that our hypotheses are correct (Popper, 1959, 1994). Theories can never be proven, confirmed, or verified; rather, if they fail repeated and determined attempts at falsification, they are said to be "robust" and more likely to be correct than those which have not been extensively tested. Scientific progress thus depends on repeated tests, or challenges, of hypotheses.

The consequence of this is that scientific thinking is not static. When data do not fit the scientific theory, a "paradigm shift," that is, a fundamental shift in thinking or even basic assumptions, must occur (Kuhn, 1970). Miller (2008) notes that science is a revolutionary activity since new discoveries can change our view of the world and overturn our accepted ideas of nature. Such was the case with the theory of evolution.

## A Very Brief History of the Theory of Organic Evolution

### Evolution (Pre-Natural Selection): 1600s–1859

While today biological (or "organic") "evolution" is commonly associated exclusively with Charles Darwin, the idea of evolution predates him considerably. Until the Enlightenment, most Europeans conceived of a static Earth. Nonetheless, by with the Enlightenment, a small minority of intellectuals, the natural philosophers, came to recognize that Earth and its organisms had indeed changed; this was referred to it as "transformism" (Sloan, 2019). Danish cleric Nicholas Stenö (1638–1686), first identified and codified a changing (non-static) Earth. Seeing layers of rocks, he inferred that successions of layers were composed of successively younger rocks, and older ones resting below the younger ones. He also reached the conclusion, not obvious at the time, that fossils, whose origins had been deeply enigmatic, were the remains of once-living organisms. These observations led him to the conclusion that recognizing that older rock layers represented environments no longer present in the places where they were observed (Cutler, 2003). Stenö's conclusions fit nicely with those of Robert Hooke (1635–1703), who, although best-known for his "discovery" of cells, also concluded that fossilized objects like petrified wood and fossil shells were the remains of once-living things that had been soaked in lithifying water laden with minerals. Brilliantly, he analogized fossils with old Roman coins: just as these could be used to date cities, the fossils could be used to date successive layers rocks. By the 1700s, George-Louis Leclerc, Comte de Buffon (1707–1788), had connected several naturalistic theories of earth history, embryological development, and changes of organic species in response to external changes of conditions of life (Sloan, 2019).

Within this intellectual ferment appeared Jean Léopold Nicolas Frédéric, Baron (Georges) Cuvier (1769–1832) who, at the height of his powers, was recognized as perhaps the world's greatest anatomist. Cuvier rejected the gradual transformation of species, at least in part because if species were mutable, without clear boundaries between them, it would be impossible for him to create an accurate taxonomy based on principles of comparative anatomy (Coleman, 1964). Instead of gradual change, Cuvier developed a theory of catastrophes. According to this idea, the fossil record shows that animal and plant species were abruptly destroyed time and again by deluges and other natural cataclysms—(i.e., they went *extinct*)—and that new species appeared only after those catastrophic events (Appel, 1987). Here Cuvier attempted to reconcile the conflict between the static Earth of biblical scholars, and the changing Earth of the natural scientists: in Cuvier's telling, each catastrophic event reset the Earth; the most recent catastrophic event being the Biblical Flood. Some people during Cuvier's lifetime could not accept the idea of extinction because it was religiously troubling. For these thinkers, God had created all of nature according to a divine plan at the beginning of the Earth, and therefore it would not make sense for God to kill off some parts of His work. If life extended from ocean slime to humans to angels, extinctions would remove vital links from the divine creation (Understanding Evolution, 2020a).

Cuvier's one-time friend and colleague, Étienne Geoffroy Saint-Hilaire (1772–1844), worked out the implications of the inner anatomical similarities among vertebrates. Using his two main principles, the "principle of connection" and the "law of balance," St. Hilaire described the unity of the animal kingdom (Appel, 1987). He also became interested in the relationships between the changes through time of lineages and the changes in organismic development (e.g., ontogeny), which foreshadowed relationship between evolutionary theory and developmental biology (evo-devo; see below). St. Hilaire also recognized successions of lineages going from ancient time until the present, believing that changes in form were abrupt and driven by the environment. This should not be confused with common descent (see below), but rather the idea that particular types held the potential to modify into other forms.

Charles Darwin's grandfather, Erasmus Darwin (1731–1802), one of the leading intellectuals of 18th-century England, was also an early evolutionist. He proposed a theory of species development based on a concept of dynamic living matter. He proposed that life evolved from a single common ancestor, forming "one living filament," and that competition and sexual selection could cause changes in species (Understanding Evolution, 2020b). Ultimately, however, missing from the elder Darwin's ideas about biological evolution was an appropriate mechanism.

In his book *Philosophie Zoologique* (1809), French naturalist and geologist Jean-Baptiste Lamarck (1744–1829) reaffirmed the refinement of biological adaptations for survival in different environments. Lamarck proposed that the evolution of animals was propelled by their needs and that these needs in turn, created behavioral changes that caused structural changes which made the behavior more efficient. Ultimately, Lamarck's idea was that the changes that an individual developed during its lifetime in response to the environmental needs were inherited by its offspring—the so-called *inheritance of acquired characters* (that is, offspring inherited characters acquired during the parents' lifetimes). Like many natural scientists of the time, Lamarck's ideas were by no means universally accepted; however, Lamarck's ideas were picked up long after his death (and even ~80 years after Darwin), with the gravest of consequences for Soviet agriculture, by Joseph Stalin's preferred biologist, T.D. Lysenko (1898–1976).

This brief *dim sum* of pre-Darwinian evolutionists would not be complete without the Darwinian contemporary Charles Lyell (1797–1875). Lyell's explanation of observable geology, with processes such as uplift, erosion, and sedimentation, required an estimated millions of years of earth history to generate the phenomena of massive mountain ranges and of deep river valleys, and this gradual timeline was also reflected in Darwin's ideas about the gradual changes in species over time. Perhaps equally important for Darwin, was Lyell's (1833) fundamental idea of a dynamic Earth, which, too, found itself reflected in the changes through time undergone by lineages of organisms (Bowler, 2003).

### Non-Evolutionist Natural Scientists

Although the ideas of evolution, changes in organisms through time, were popular by the time Darwin wrote his famous book, there were scientists who did not agree with these concepts. Perhaps the most famous scientific anti-evolutionist was Richard Owen (1804–1892). Owen's own work built on the French anatomical work of Cuvier and St. Hillaire and he introduced many of the terms still used today in anatomy and evolutionary biology, including the words "dinosaur" and "homology" (the idea that two structures in two different organisms can be traced back to a single structure in the most recent common ancestor of those two organisms). Using homology, Owen reasoned that there must be a common "archetype" or structural plan for all vertebrates. Today we know that this is the case, however, Owen did not believe in a common ancestor to all the vertebrates; rather Owen thought that the archetype represented an idea in the Divine mind, a deity which already knew all the possible modifications *à la* St. Hillaire (Cadbury, 2000; Desmond, 1986).

## Darwin's Contribution

### Evolution by Natural Selection: 1859–Present

Charles Darwin (1809–1882) sent shockwaves around the world in 1859 with the publication of the *On the Origin of Species*. But as we have seen, the raw material for Darwin's theory had been explored for almost a full century. Geologists and paleontologists had provided evidence for the case that life had been on Earth for a long time, that it had changed over that time, and that many species had become extinct (Bowler, 2003). Darwin's genius, then, was to show how all this evidence pointed to the evolution of species from a common ancestor and to offer *natural selection* as the plausible mechanism by which that evolution might take place.[2] As summarized in Fastovsky and Weishampel (2021) Darwin constructed the following well-known argument in *On the origin of Species*:

1. *Domesticated animals and plants show a wide range of variation.*
2. *A wide range of variation also exists among wild animals and plants as well.*
3. *All living creatures are engaged in a "struggle" (Darwin's terminology) to survive and ultimately reproduce, and that struggle is most severe among those individuals that are most closely related.*
4. *The struggle to survive in combination with the variation that exists among individuals leads to the survival and, most importantly, production of viable progeny in some variants as opposed to some others, a process that Darwin called natural selection.*
5. *The viable progeny of some variants as opposed to others ensures that the characteristics of the successfully reproducing variants make it into the next generation.*
6. *This process, repeated over hundreds or even thousands of generations, is evolution by natural selection, sometimes called "Darwinian evolution" (p. 57).*

Darwin's book was a best-seller and highly controversial. Its hypothesis was a-theistic; God seemed neither to be guiding the process, nor even to be involved. Moreover, it suggested that disturbingly large amounts of time were required for the changes proposed. But how much time?

## Evolutionary Theory after Darwin

### The Genetics Revolution and Mutationist School of Evolution: ~1910–1920s

Darwin was unaware of the genetic basis of inheritance, although the famous genetic studies of peas by Austrian monk Gregor Johann Mendel (1822–1884) were available to him. But as genetics became better understood and was incorporated into the biological opus, a group of scientists called the mutationists used Mendel's ideas to explore evolution; these included Hugo De Vries (1848–1935), William Bateson (1861–1926), Thomas Hunt Morgan (1866–1945), and Richard Goldschmidt (1878–1958). Unlike Darwin, these scientists supported discontinuous evolution rather than gradual changes, as they were convinced that new species and varieties appeared abruptly, in leaps caused by mutations (Sapp, 2003), as the fossil record seemed to show.

To most biologists, however, who had originally turned to Mendelian genetics and championed it in the hope that it would be the basis for evolution by mutations, it seemed clear that gene changes, generally producing deleterious effects, were not the stuff of evolution after all. Genetics had to be involved, but how did this translate into evolution?

### New/Modern Synthesis: Late 1920s–Present

The "new" or "modern synthesis" came from the wildly successful collaboration, beginning in the late 1920s, of a number of scientists, primarily biologists, to merge their narrow, individual specializations to form an encompassing synthetic theory of evolutionary biology. These early pioneers included J.B.S. Haldane (1892–1964; genetics and biostatistics), Ernst Mayr (1904–2005; population biology and ornithology), Theodosius Dobzhansky (1900–1975; genetics), George Gaylord Simpson (1902–1984; paleontology), Sewell Wright (1889–1988; population genetics), George Ledyard Stebbins Jr. (1906–2000; botany), and Ronald Fisher (1890–1962; biostatistics). The New Synthesis of evolutionary theory differed from the mutationists' ideas by the conclusion that evolution did not occur by large, discontinuous, macromutational changes, but instead by the gradual accumulation of traits with selective value (Sapp, 2003). A quantitative, evolving model of evolution was developed, that was predictive of what has been observed in nature ever since the scientists above, and their academic descendants, first collaborated. Thus, the New Synthesis became the dominant model for evolutionary biology for the three—four generations of evolutionary biologists who succeeded them. Indeed, it is still almost exclusively what is taught in high schools, and while it has been neither falsified nor rejected, the "tempo and mode" (to quote G.G. Simpson, 1944) of evolution as well as many other aspects of the process, are not completely explained by its principles.

### Punctuated Equilibrium/Macroevolution (Echoes of the Old Mutationist School): 1972–Present

Among the observations not well explained by the modern synthesis was the abrupt appearance of new forms in the fossil record. The evolutionary story dating from Darwin (who had been influenced by Lyell), was that development of new species occurred via the gradual accumulation of small variations over long periods of time until the descendent was sufficiently different from its ancestor that it merited the designation of being a new species. So it was that Darwin—and other successive generations of scientists—interpreted the origin of the diversity seen in the modern finch populations of the Galapagos Islands (e.g., "Darwin's finches"). However, paleontologists Niles Eldredge (1943–) and Stephen Jay Gould (1941–2002) proposed that, empirically, this is not what the evolutionary record, as revealed in fossils, reveals. They therefore read the fossil record literally: species are generally stable, changing little for 1–2 million years, but can encounter a "punctuated" or rapid burst of change that results in a new species (Eldredge & Gould, 1972). Their model, which they termed "punctuated equilibrium," proposed that species emerged in a very short geological period of time followed by long periods of stasis. The model was radical (paralleling societal changes of the time), sounded mutationist overtones, and required a mechanism for large anatomical changes in very brief amounts of time. Evolutionary biologists, functioning within the paradigm of the "New Synthesis" could not make it work comfortably.

### Evolution/Development (Evo-Devo): 1990s–Present

The solution began to materialize in the early 1980s when scientists revisited an important relationship: that between evolution and development (nicknamed "evo-devo"). The ideas encompassed by evo/devo can be traced back to Ernst von Baer (1792–1876) who noticed that during embryological development, individual embryos sequentially added physical features, so that the first features to appear were those that appeared most widespread in a particular phylum, and as development occurred, the features that appeared became more and more specialized. Von Baer did not consider these observations as having implications for evolution and saw developmental progression of the embryo as simply paralleling the taxonomic order of life from simple to complex (Sapp, 2003).

However, to the German Ernst Haeckel (1834–1919), von Baer's observations could be interpreted evolutionarily. Haeckel believed that a developing embryo, from fertilization to gestation or hatching (i.e., during ontogeny), goes through stages representing successive adult ancestors in the evolution of the animal (phylogeny), and from this came the famous expression, "ontogeny recapitulates phylogeny."[3]

Gould (1977) wrote a seminal book on the subject ("Ontogeny and Phylogeny"), and ultimately it became clear from the work of many scientists that one way to effect large changes in very short amounts of time was through modifications in the rate of expression of the so-called "regulatory" genes, which control the timing of the expression of "structural" genes, which code for particular structures (limbs; hair color, etc.). These kinds of modifications in the timing of gene expression were called *heterochrony* (McKinney & McNamara, 1991; McNamara, 2012), and neatly solved the problem of how to make big changes rapidly without damaging the integrated functionality of the organism. Heterochrony allowed for substantial changes using the timing of development, while maintaining the fundamental integrity of the developmental (ontogenetic) process. Understanding heterochrony has had profound ramifications for our understanding of homology, and organic relationships.

## A Modern View of the Theory of Evolution by Natural Selection

A modern understanding of evolution by natural selection is a rich pastiche of this long intellectual history. Darwin's basic contribution, the process of natural selection, is still as valid as it was when Darwin (and Wallace) first proposed it; yet, the field has moved far beyond Darwin's model. So far, in fact, that biologist and New York Times columnist Carl Safina wrote a column entitled, "Darwinism must die so that evolution may live:"

> *Equating evolution with Charles Darwin ignores 150 years of discoveries, including most of what scientists understand about evolution. Such as: Gregor Mendel's patterns of heredity (which gave Darwin's idea of natural selection a mechanism—genetics—by which it could work); the discovery of DNA (which gave genetics a mechanism and lets us see evolutionary lineages); developmental biology (which gives DNA a mechanism); studies documenting evolution in nature (which converted the hypothetical to observable fact); evolution's role in medicine and disease (bringing immediate relevance to the topic); and more.*
>
> (Safina, 2009)

The fact is that the theory of evolution is one of the best-supported theories in the history of science by evidence from a wide variety of scientific disciplines, including, but not exclusively, those marshaled to understand it during the New Synthesis. While it is not within the scope of this chapter to detail how well supported the theory of evolution by natural selection actually is, readers are referred to Coyne (2010) for a comprehensive but readable treatment of this subject.

## Time

### Numerical Age of Earth

All of this would be well and good but as the late geochronologist Samuel Bowring noted "No dates; no rates." If you can't date things, you can't figure out how quickly (or slowly) anything takes place. And thus, as humans first began to document events and investigate the natural world, a truly fundamental question was the age of the Earth, since this knowledge obviously had implications about how rapid the changes on Earth (including its biota) took place. Aristotle (c. 385 BC – 323 BC) thought the Earth had always existed, while Roman poet Lucretius (99 BC – c. 55 BC) posited that the Earth's formation must have been relatively recent, since there were no records going back beyond the Trojan War (Braterman, 2013). Famously in 1650, Archbishop James Ussher (1581–1656) of Ireland used a biblical chronology[4] to offer the date of October 23, 4004 BC from the creation of Earth to the birth of Christ. By 1760, De Buffon estimated the age of the Earth to be 75,000 years.

A variety of natural scientists, following Lyell's studies of the rates at which sedimentary rocks were deposited and eroded, produced estimates of the age of the Earth in millions of years, but even though some rock sequences (and the fossils that they contained) were thought to be older, numerical dates for the age of the Earth remained elusive (Dott & Batten, 1971).

All that dramatically changed in the latter part of the 19th century, when William Thomson, Lord Kelvin (1824–1907), calculated a numerical age for the Earth. He asserted that the earth was a red-hot globe that has cooled over millions of years since its formation. By the end of the 19th century and after some earlier waffling, Kelvin settled on his calculation that the Earth was 20–40 million years old. As was only later understood, Lord Kelvin's date(s) grossly underestimated the Earth's antiquity, because radioactive materials in the core generate heat via decay.

By 1904, Ernest Rutherford (1871–1937) and his colleagues figured out how to date rocks (and thereby invented the science of *geochronology*, that is, determining the ages of Earth materials) using radioactive-decay products and he famously made the case for an even older Earth at a lecture attended by Lord Kelvin himself (Whitehouse, 2016). By 1927, geologist Arthur Holmes (1890–1965) had published *The Age of the Earth, an Introduction to Geological Ideas* in which he presented a range of 1.6 to 3.0 billion years for the age of the earth (Dalrymple, 1994). Consensus for the age of the Earth has been reached finally after many tests on independent data sets—which of course is just the way in which science works. Darwin understood that the Earth's extreme antiquity would be required for his proposed method for biological evolution, and the current age of the Earth, estimated to be 4.55 ± 0.05 billion years (Braterman, 2013), is completely concordant with Darwin's ideas although, perhaps, beyond even his imagining.

## Controversies and Attacks on Evolutionary Theory

Given the evident scientific support for evolution by natural selection, it may be surprising that more than a century after Darwin's death in 1882, there is still skepticism about the validity of his contributions (and those of generations of his successors). The mere fact that a variety of scientific analyses, using many different indicators, have failed to falsify the hypothesis of evolution by natural selection (largely as proposed by Darwin and Wallace), and that evolution is unquestionably the unifying concept of the biological sciences, generates the obligation to teach it to students. Furthermore, the concepts required to understand evolution by natural selection like evolutionary change, genetic variation, and the struggle for existence, are not particularly hard to grasp because at an introductory level they do not have the mathematical complexities that impede understanding of major theories in physics such as relativity and quantum mechanics (Thagard & Findlay, 2009). So at least in part, the controversies surrounding the theory of evolution must be related to misunderstanding what the theory actually says or, perhaps more significantly, how the theory is seen by some to conflict with theistic perceptions of Earth time and biotic diversity.

Since its early inception, evolutionary theory has been attacked; the most famous debate during Darwin's time taking place on June 30, 1860, between biologist Thomas Huxley and Oxford Bishop Samuel Wilberforce (Hesketh, 2009). In the United States, people have challenged the legality of teaching evolutionary theory in public schools. Perhaps the most famous case about evolution in the United States was the 1925 Scopes trial, which prosecuted John Scopes, a football coach and substitute biology teacher in Tennessee, for teaching a theory of human origins that contradicted "the Divine creation of man as taught in the Bible" (Miller, 2008, p. 89). Since that time, many additional cases of importance have been tried by the courts, including 1968, Epperson v. Arkansas; 1981, Segraves v. State of California; 1982, McLean v. Arkansas Board of Education; 1987, Edwards v. Aguillard; 1990, Webster v. New Lenox School District; 1994, Peloza v. Capistrano School District; 1997, Freiler v. Tangipahoa Parish Board of Education; 2000, Rodney LeVake v Independent School District 656, et al.; 2005, Selman et al. v. Cobb County School District et al.; and most recently, December 2005, Kitzmiller et al. v. Dover Area School District (Matsumura & Mead, 2016). These court decisions have all ruled that the various forms of religious ideas, including intelligent design, and creationism, are belief systems (e.g., religion), not science, and that it is therefore unconstitutional to include them in public school science classes. Evolutionary theory remains the only scientifically valid explanation for the history of life on Earth, while religious beliefs, whatever else is truth, are not science.

One of the most widespread oppositions to evolutionary theory arises from its contravening the idea of divine creation. However, this discordance may be based on ignorance of both evolutionary theory and religion. For example, the Roman Catholic Church, the largest Christian church, has publicly stated that teachings on creation could coexist with evolutionary theory, when Pope Pius XII wrote the encyclical "*Humani Generis*" (1950). In this document, the Catholic Church distinguished between the *spiritual*, which is the domain of God, and the *physical*, which is the domain of science. It is a distinction that carries a lot of weight with students when it is explained. The spiritual, not being science, does not have the same methods as science. The document states,

> *The teaching authority of the Church does not forbid that, in conformity with the present state of human sciences and sacred theology, research and discussions, on the part of men experienced in both fields, take place with regard to the doctrine of evolution, in as far as it inquires into the origin of the human body as coming from pre-existent and living matter—for the Catholic faith obliges us to hold that souls are immediately created by God.*
>
> (Catholic Church, 1950, # 36)

Even more supportive of evolution, Pope John Paul II (1996), on his "Message to the Pontifical Academy of Sciences: On Evolution" called evolution "more than a hypothesis." More recently, Pope Francis stated, "God is not a demiurge or a magician, but the Creator who brought everything to life. . . . Evolution in nature is not inconsistent with the notion of creation, because evolution requires the creation of beings that evolve" (Welsh, 2014).

Likewise, according to the Pew Research Center (2014), other religions have stated that there are alignments between religion and evolutionary theory. For example, Buddhist scholars have described that there is consistency between evolution and Buddhism, such as the notion that all life is impermanent. The teachings of other religious groups including the Church of Jesus Christ of Latter-day Saints, the Episcopal Church, the Evangelical Lutheran Church in America, the Presbyterian Church, the United Methodist Church, and Hinduism, do not generally find an inherent conflict between evolutionary theory and their faith (Pew Research Center, 2014).

Thagard and Findlay (2009) highlight two main obstacles to acceptance of the scientific explanation for the history of life: cognitive and emotional. They propose that cognitive obstacles can arise in the form of conceptual difficulties such as the fact that evolutionary theory involves probabilistic changes in genes and species, the so-called "population thinking" of Mayr (1982). Since evolution happens at the population level rather than the individual level, some students may be drawn to Lamarckian explanations in terms of the heritability of acquired characteristics, which is most like the rapid, directed changes that take place in individual's personal lives and cultural developments.

Perhaps most often mentioned, is the fact that some religious traditions such as Judaism, Christianity, and Islam assert divine creation as the origin of species, rather than evolution by natural selection. The Bible presents a definitively different story from the biological view that organisms on earth evolved over the course of billions of years. Therefore, some students who are literal interpreters of these texts need to either reject it or revise their views dramatically before they become capable of accepting evolutionary explanations of the origins of species.

Moreover, due to emotional reactions and the appealing aspects of the religious-psychological perspective that God formed human souls as the culminating act of creation, some people would prefer that the theory of evolution be false (Thagard & Findlay, 2009). The idea of a loving and all-powerful God may provide some people reassurance that the many problems that life presents are just transitory parts of a beneficent, divine plan, which will eventually lead to everlasting happiness. Also, if humans are free agents, as we might infer from our apparent abilities to make choices, Darwinian approaches to biology may make some think that their lives have been rendered meaningless, devoid of spirituality, purpose, and morality; lives in effect reduced in value (e.g., Harms & Reiss, 2019). Because religion and evolution are thought to be incompatible by many people in the public eye, including some dogmatic religious leaders and politicians (Barnes et al., 2017), students may perceive that evolution is in conflict with religious beliefs.

## Evolution in the Public Eye

Recent research continues to show that an understanding of evolutionary theory is poor among school students and the general population alike (Reiss &

Harms, 2019). Some of the misunderstandings come from doctrinaire religious beliefs, while others come from ignorance about what the theory is and how it is supported. Some students reject the theory of evolution because they erroneously understand it to be incompatible with their religious beliefs. Some researchers have argued that abandoning religious belief is the only way to help religious individuals accept evolution (e.g., Lawson & Weser, 1990; Honey, 2015). Staunch atheistic evolutionists like Oxford University professor and prolific author Richard Dawkins argue that science can provide an explanation for everything in the observable universe, and thus science suggests that a supernatural creator almost certainly does not exist (Dawkins, 2008). Dawkins (1989), goes as far as writing, "It is absolutely safe to say that if you meet somebody who claims not to believe in evolution, that person is ignorant, stupid or insane" (p. 34).[5] Publicly insulting people may be counterproductive to helping people to understand the issues or feel invited to consider different points of view (Winberg, 2017). Alternatively, some researchers have found that highlighting the faith/evolution compatibility is an effective means to increase student acceptance of evolutionary theory (Miller, 2008; Lindsay et al., 2019; Barnes et al., 2017).

Finally, another potential source of confusion underlying misunderstandings of the theory of evolution is related to language. Many words in science, like "adaptation," "fitness," and "theory," also appear in everyday language with (sometimes) subtly different meanings (Sinatra et al., 2008). Therefore, teaching evolution requires a broad explanatory palette as well as a deft sensitivity to peoples' concerns and fears, to address misconceptions and misunderstandings.

## Conclusion

If we do not offer a rigorous introduction to evolutionary theory or if we compromise its treatment, we deprive students of the coalescing scientific concept of the biological sciences. In short, Biology cannot be properly taught without teaching the theory of evolution with all of its predictive and explanatory powers.

Studying evolution is an excellent conduit for allowing students to learn about the process of scientific inquiry, because as we have seen, evolution offers myriad and diverse examples of the ways that scientists gather and analyze information, test competing hypotheses, and come to a consensus about explanations for natural phenomena. Indeed, Belin and Kisida (2015) report that in the United States, there is a clear and consistent relationship between a state's public acceptance of evolution and the levels of science achievement among students in that state. Understanding science is essential for making informed decisions and participating in a global society (Calabrese Barton & Upadhyay, 2010). Limitations on exposure to evolutionary theory in the science classroom compromise understanding of a fundamental tenet of biology and medicine, and also undermine of how scientific knowledge is interpreted. It is thus essential that all students be offered a complete science education that includes a full and correct introduction to evolutionary biology.

Finally, as will be explored further in this text, if learned properly, evolutionary biology has the potential to help students apply their insights beyond biology. From Chapters 2 to 4, we demonstrate that evolutionary theory as a venue to explore social justice works both in terms of science education at the high-school level and personal empowerment. No other science can make that claim, so it is evolutionary theory that is best suited to unite biology, to empower by science education, and to underpin social justice. This text, then, contains a detailed, scholarly exploration of the unique and largely unexplored relationship between evolution biology and social justice.

*The nuns, my aunts, introduced me to the exploration and thrill of learning about the natural world, but it was not until college that I first learned about the theory of evolution by natural selection—this would be the idea that coalesced my thinking about the biological world and our place in it. As an adult, I came to have the honor of sharing with those nuns what I had learned in college, describing for them my passion for the discovery of the living things, and celebrating the power that science had given me and my family. For them, it confirmed what they already believed—humans are intrinsically connected to every living thing, to the Earth itself, and to the events of the cosmos. For me, evolutionary theory was the crux of my liberation from poverty and scientific ignorance.*

## Notes

1 A recent example of this was the 2020 Coronavirus pandemic, in which the science was unambiguous about minimizing disease transfer; yet, largely for economic and political (non-scientific) reasons some American politicians wanted to open venues and encourage behaviors that largely facilitated viral spread. Their non-scientific, subjective decisions contributed to >800,000 deaths in the United States as of 01/22 (now, over 1.2 million deaths in the United States; however, this can no longer be laid at the feet of irresponsible politicians), and, at the height of the pandemic, the highest rates of infection in the world.

2 Similar ideas about evolution were developed simultaneously by Alfred Russel Wallace (1823–1913). Wallace considered Darwin a scientific mentor and in 1858 wrote to him from the field in Indonesia proposing the idea that he'd had: that of evolution by natural selection. Darwin was, naturally, aghast to find that the theory that he had been developing for 20+ years was invented in parallel by somebody else. Friends intervened, and it was agreed that the two of them jointly read a paper to the Linnaean Society in 1858, arguing the theory of evolution by natural selection.
3 Haeckel's precise formulation was ultimately abandoned by evolutionary biologists who recognized that embryos pass through stages that often resemble ancestral embryonic types, but not *adult* types.
4 *Annales Veteris Testamenti* (1650)
5 An unintended irony by Professor Dawkins, since evolution, as we have seen, is not something one "believes in;" it is a scientific hypothesis (and a well-supported one at that).

## References Cited

Appel, T.A. (1987). *The Cuvier-Geoffroy Debate: French Biology in the Decades Before Darwin*. Oxford: Oxford University Press.

Barnes, M.E., Elser, J., & Brownell, S.E. (2017). Impact of a Short Evolution Module on Students' Perceived Conflict between Religion and Evolution. *The American Biology Teacher*, *79*(2), 104–111.

Belin, C.M., & Kisida, B. (2015). State Science Standards, Science Achievement, and Attitudes About Evolution. *Education Policy*, *29*(7), 1053–1075.

Bowler, P.J. (2003). *Evolution: The History of an Idea*, 3rd ed. Berkeley, CA: University of California Press.

Braterman, P. (2013, October 20). *How Science Figured Out the Age of Earth*. Retrieved July 9, 2020, from www.scientificamerican.com/article/how-science-figured-out-the-age-of-the-earth/

Cadbury, D. (2000). *The Dinosaur Hunters: A True Story of Scientific Rivalry and the Discovery of the Prehistoric World*. London: Fourth Estate.

Calabrese Barton, A., & Upadhyay, B. (2010). Teaching and Learning Science for Social Justice: Introduction to the Special Issue. *Equity & Excellence in Education*, *43*(1), 1–5.

Catholic Church, & Fenton, J.C. (1950). *Humani Generis: Encyclical Letter of Pope Pius XII Concerning Some False Opinions Which Threaten to Undermine the Foundations of Catholic Doctrine*. www.vatican.va/content/pius-xii/en/encyclicals/documents/hf_p-xii_enc_12081950_humani-generis.html

Coleman, W. (1964). *Georges Cuvier, Zoologist a Study in the History of Evolution Theory*. Cambridge, MA: Harvard University Press.
Coyne, J. (2010). *Why Evolution Is True*. New York: Penguin Books.
Cutler, A. (2003). *The Seashell on the Mountaintop: A Story of Science, Sainthood, and the Humble Genius who Discovered a New History of the Earth*. Cambridge: Cambridge University Press.
Dalrymple, G.B. (1994). *The Age of the Earth*. Stanford, CA: Stanford University Press.
Dawkins, R. (1989). Book Review of Donald Johanson and Maitland Edey's Blueprint. *New York Times*, section 7, 34.
Dawkins, R. (2008). *The God Delusion*. New York: Mariner Books.
Desmond, A. (1986). *Archetypes and Ancestors*. Chicago, IL: University of Chicago Press.
Dobzhansky, T. (1973). Nothing in Biology Makes Sense Except in the Light of Evolution. *American Biology Teacher*, *35*(3), 125–129.
Dott, R.H., & Batten, R.L. (1971). *Evolution of the Earth*. New York: McGraw-Hill Book Co.
Eldredge, N., & Gould, S.J. (1972). Punctuated Equilibria: An Alternative to Phyletic Gradualism. In T.J.M. Schopf (Ed.), *Models in Paleobiology*. San Francisco: Freeman Cooper.
Fastovsky, D., & Weishampel, D. (2021). *Dinosaurs. A Concise Natural History*. Cambridge: Cambridge University Press.
Gould, S.J. (1977). *Ontogeny and Phylogeny*. Cambridge, MA: Belknap Press.
Harms, U., & Reiss, M.J. (2019). The Present Status of Evolution Education. In U. Harms & M.J. Reiss (Eds.), *Evolution Education Re-Considered: Understanding What Works*, pp. 1–19. Cham, Switzerland: Springer Nature.
Hesketh, I. (2009). *Of Apes and Ancestors: Evolution, Christianity, and the Oxford Debate*. Toronto, Canada: University of Toronto Press, ISBN 9780802092847.
Honey, L.P. (2015). Why I Teach the Controversy: Using Creationism to Teach Critical Thinking. *Frontiers in Psychology*, *6*(793), 1–5.
Kitzmiller v. Dover area school dist., 400 F. supp. 2d 707 (M.D. Pa. 2005). Justia Law. (n.d.). Retrieved April 21, 2022, from https://law.justia.com/cases/federal/district-courts/FSupp2/400/707/2414073/
Kuhn, T.S. (1970). *The Structure of Scientific Revolutions*. Chicago: University of Chicago Press.
Lamarck, J. (1809). *Philosophie zoologique by Jean Baptiste Pierre Antoine de Monet de Lamarck*. Retrieved June 22, 2020, from www.cambridge.org/core/books/philosophie-zoologique/CC98F14B9869695D634C9D2DDF3F5927
Lawson, A.E., & Weser, J. (1990). The Rejection of Non-Scientific Beliefs About Life: Effects of Instruction and Reasoning Skills. *Journal of Research in Science Teaching*, *27*, 589–606.

Lindsay, J., Arok, A., Bybee, S.M., Cho, W., Maskiewicz Cordero, A., Ferguson, D., Galante, L., Gill, R., Mann, M., Peck, S., Shively, C., Stark, M., Stowers, J., Tenneson, M., Tolman, E., Wayment, T., & Jensen, J. (2019). Using a Reconciliation Module Leads to Large Gains in Evolution Acceptance. *CBE Life Sciences Education, 18*(ar58), 1–11.

Lyell, C. (1833). *Principles of Geology*. Retrieved June 22, 2020, from http://lhldigital.lindahall.org/cdm/ref/collection/earththeory/id/17524

Matsumura, M., & Mead, L. (2016, June). *Ten Major Court Cases About Evolution and Creationism: National Center for Science Education*. Retrieved June 25, 2020, from https://ncse.ngo/ten-major-court-cases-about-evolution-and-creationism

Mayr, E. (1982). *The Growth of Biological Thought*. Cambridge, MA: Harvard University Press.

McKinney, K.L., & McNamara, K.J. (1991). *Heterochrony: The Evolution of Ontogeny*. New York: Springer.

McNamara, K.J. (2012). Heterochrony: The Evolution of Development. *Evolution: Education and Outreach, 5*, 203–218. https://doi.org/10.1007/s12052-012-0420-3

Miller, K.R. (2008). *Only a Theory: Evolution and the Battle for America's Soul*. New York: Penguin Books.

Pew Research Center. (2014). *Religious Groups' Views on Evolution*. Retrieved June 25, 2020, from www.pewforum.org/2009/02/04/religious-groups-views-on-evolution/

Pope John Paul II. (1996, October). *Message to the Pontifical Academy of Sciences: On Evolution*. Retrieved June 25, 2020, from https://humanorigins.si.edu/sites/default/files/MESSAGE%20TO%20THE%20PONTIFICAL%20ACADEMY%20OF%20SCIENCES%20(Pope%20John%20Paul%20II).pdf

Pope Pius XII. (1950, August). *Humani Generis (August 12, 1950): PIUS XII*. Retrieved June 25, 2020, from www.vatican.va/content/pius-xii/en/encyclicals/documents/hf_p-xii_enc_12081950_humani-generis.html

Popper, K.R. (1959). *The Logic of Scientific Discovery*. New York: Basic Books. A translation by the author with the assistance of Julius Freed and Jan Freed of *Der Logik der Forschung*. Vienna: J. Springer, 1935.

Popper, K.R. (1994). *All Life Is Problem Solving*. London: Routledge.

Reiss, M.J., & Harms, U. (2019). What Now for Evolution Education? In U. Harms & M.J. Reiss (Eds.), *Evolution Education Re-Considered: Understanding What Works*, pp. 331–343. Cham, Switzerland: Springer Nature.

Safina, C. (2009, February 9). Essay: Darwinism Must Die so That Evolution May Live. *New York Times*.

Sagan, C. (1997). *The Demon-Haunted World: Science as a Candle in the Dark*. London: Headline Book Publishing.

Sapp, J. (2003). *Genesis: The Evolution of Biology*. New York: Oxford University Press.

Simpson, G.G. (1944). *Tempo and Mode in Evolution*. New York: Columbia University Press.

Sinatra, G.M., Brem, S.K., & Evans, E.M. (2008). Changing Minds? Implications of Conceptual Change for Teaching and Learning about Biological Evolution. *Evolution: Education and Outreach, 1*(2), 189–195. https://doi.org/10.1007/s12052-008-0037-8

Sloan, P. (2019). Evolutionary Thought Before Darwin. In E.N. Zalta (Ed.), *The Stanford Encyclopedia of Philosophy Edward*. https://plato.stanford.edu/archives/win2019/entries/evolution-before-darwin/

Thagard, P., & Findlay, S. (2009). Getting to Darwin: Obstacles to Accepting Evolution by Natural Selection. *Science & Education, 19*, 625–636.

Understanding Evolution. (2020a). *Extinctions: Georges Cuvier*. University of California Museum of Paleontology. Retrieved February 12, 2023, from https://evolution.berkeley.edu/evolibrary/article/0_0_0/history_08

Understanding Evolution. (2020b). *Erasmus Darwin*. Retrieved February 12, 2023, from https://ucmp.berkeley.edu/history/Edarwin.html

Welsh, T. (2014, October 28). Pope Francis Says Science and Faith Aren't at Odds. *U.S. News & World Report*. Retrieved July 19, 2020, from www.usnews.com/news/articles/2014/10/28/pope-francis-comments-on-evolution-and-the-catholic-church

Whitehouse, D. (2016). *Into the Heart of Our World*. New York: Pegasus Books.

Winberg, O. (2017). Insult Politics: Donald Trump, Right-Wing Populism, and Incendiary Language. *European Journal of American Studies, 12*(2), document 4, Online since August 2, 2017, connection on June 26, 2020. http://journals.openedition.org/ejas/12132

# Part IV

## The Origins of Racism and its Relationship to Science in America

# 8

# The Roots of Oppression

## Slavery, Colonization, and Unfettered Free Markets

*When I was in second grade, I tore into pieces the "Best Student Award" certificate from the school* Liceo Salazar y Herrera. *My mother was furious and could not begin to understand why I would "misbehave" in this way. She was proud of her youngest son, and this certificate symbolized not just my own work, but her efforts and dedication as a struggling single mother. It was only after she noticed the tears running down my face that she asked why I had ripped the paper. I told her that I did not feel that I deserved the award; there was another student, Carlos, that I thought worked even harder than I did—and as poor as we were, he was even poorer. Many Latin Americans—like Carlos and me—tend to be mixed—the descendants of bastard children of the Spaniards who procreated with the indigenous peoples they conquered and suppressed on the one hand, and the Africans they brought to the New World as slaves on the other. Unlike Carlos whose physical characteristics were associated with the native Chibcha peoples, I looked like our second-grade teacher, with a heavier dose of Mediterranean coloration. Aside from the certificate, the award came with the choice of seats, an invitation to sit on the stage during a school-wide ceremony, and the option to join the marching band. None of these things were offered to Carlos. Because of the award, my world opened up in school, and even discipline was dispensed out differently for me. Carlos and I had been differentially treated as a result of the color of our skin. The teacher was responding consciously or unconsciously to her perception of our respective races and their ranking in her system of values; in short, her response was racist.*

To understand, let alone teach, the evils of racialized oppression such has been carried out globally by western nations over the past ~400 years, a solid

DOI: 10.4324/9781003409144-12

background in the history of this kind of oppression is required. In this and the following chapters, therefore, we delve into its historical evolution. What will be strikingly clear is that, virtually since the beginning, the uncomfortable fact of the matter is that science, as it was understood by its practitioners at the time, was always called upon to justify and legitimize oppression.

Caste-based social injustices appear to have been present throughout all documented history including state-sanctioned slavery of prisoners of war and overt oppression in pre-Columbian civilizations. Romans famously enslaved their defeated enemies, a slavery based on war captivity rather than race (Joshel, 2019). However, current social inequities, such as those documented in Chapter 5, are directly rooted in the history of a very different and *much* more recent race-based subjugation. This chapter will focus on this history: slavery, colonization, and free markets, and trace how it came to be that race became a chief constituent of American life and ultimately infused a caste system in a country where, in the second paragraph of its seminal document, it proclaims "We hold these truths to be self-evident, that all men are created equal, that they are endowed by their Creator with certain unalienable Rights, that among these are Life, Liberty and the pursuit of Happiness."

## A Brief Introduction to "race."

The concept of race is dynamic, as indeed Feagin (2006) notes the fluid construction of race by saying, "The category of 'race' symbolizes social conflicts by referring to human physical characteristics, yet it is not fixed, but rather variable over time." (p. 6). Kendi (2019, p. 35) saw race as "a power construct . . . that lives socially." Wilkerson (2020) considers it to be a color-coded indicator of caste (see below). Fredrickson (2002) sees race as having

> *two components: difference and power. It originates from a mindset that regards "them" as different from "us" in ways that are permanent and unbridgeable. This sense of difference provides a motive or rationale for using our power advantage to treat the. . . . Other in ways that we would regard as cruel or unjust if applied to members of our own group.*
>
> (p. 9)

At one point, race had a religious underpinning relating to the biblical Ham story (see Fredrickson, 2002; and below), but it quickly acquired a presumed biological underpinning by the time of the Enlightenment, such that skin color became an indicator of degrees of humanity with the highest degree (most human) accorded to the lightest coloration; and from then on, its application as a tool for differentiating and oppressing individuals was little changed. What has shape-shifted are the demarcations of races, and the different "science" that has been marshaled to support each conception.

Regardless of putative scientific underpinning, race has real consequences for peoples' lives. The first Black Presbyterian Church in America General Assembly moderator and Pastor Irwyn Ince Jr. (2020) said it precisely: "The idea of races is a human invention. It is a social construct, but it has a very real physical impact on the lives of people." Although as we show in Chapters 7 and 10, biology does not provide evidence for such socially-constructed distinctions, race has real implications for people's lives and it has played a major role in determining life outcomes. Our history is full of examples about the evils committed in the name of race, as we see in the following pages.

## Race-Based Enslavement

The mass enslavement of Africans can be traced back to Portuguese traders who first sailed along the Atlantic African coast in the early 1400s in search of gold. However, in their travels the Portuguese discovered the profitable business of kidnapping and trading of African people, and by 1500, over 81,000 enslaved Africans had been sold by the Portuguese and shipped to Europe, nearby Atlantic islands, and to Muslim merchants in Africa (Boddy-Evans, 2020). This, then, was a dramatic innovation: slavery had been historically tied to spoils of war, but now it became an end in and of itself, a commodity to be traded almost as valuable as the gold that originally brought the Portuguese there.

Prince Henry the Navigator (1394–1460), brother and uncle of Portuguese kings, was the first notable European to promote the idea of exclusively trading African people. The earliest expeditions sponsored by Prince Henry around the west coast of Africa were intended to be a means of avoiding Muslim traders with whom he was competing with for commerce, including the trading of humans. Ultimately acquiring human capital became the reason for sponsoring Portuguese voyages to West Africa (Kendi, 2017). In short, Prince Henry had begun a lucrative business that singled out skin coloration[1] as the key definable criterion for slavery. Prince Henry's most potent legacy might be described as the invention of race-based slavery.

After nearly two decades of lucrative slave trading led by Prince Henry, King Afonso V (1432–1481) of Portugal needed to justify the commerce of African human lives due to emerging moral and religious objections. If Africans could be understood to be somehow more bestial than Europeans, then treating them accordingly like beasts was not unconscionable. The royal chronicler Gomes Eanes de Zurara (c. 1410—c. 1474) obligingly depicted and lumped together the diverse cultures, languages, and genetics of all Africans as a single Black People who were violent and dangerous; the dark embodiment of evil (Kendi, 2017). In *The Chronicle of the Discovery and Conquest of Guinea* (1453), Zurara presents a world view in which Africans were subhuman. Zurara wrote that they were "like beasts, without any custom of reasonable beings. . . . They had no understanding of good, but only knew how to live in a bestial sloth" (Zurara, 2010, p. 84).

## Religious Justification and Support of Slavery

The Catholic Church, sensitive to opportunity, played an active role in legitimizing the slavery of blacks and indigenous peoples. The papal bull *Dum diversas*, issued by Pope Nicholas V (1397–1455) to King Alfonso V, gave the Portuguese monarchy the right to "invade, capture, vanquish, and subdue all Saracens, pagans, and other enemies of Christ, to put them into perpetual slavery, and to take away all their possessions and property" (as quoted in Newcomb, 2008, p. 98).

Slavery could also be justified by the benefits of conversion to Christianity, thus becoming a presumptive evangelical mission to the world. Many European slavers who considered themselves Christian comfortably accommodated their activities by the rationalization that slavery beneficently Christianized and civilized their captives.

For some devout Christians, however, if the slaving trade was driven by missionary motives, it then required Biblical sanction. The curse of Ham in Genesis chapter 9 verses 20 to 25 provided a suitable rationale. These verses about Noah, and his three sons, Shem, Ham, and Japheth, describe how Ham's son Canaan was cursed to become a servant because his father had seen Noah naked (see Wilkerson, 2020). Ham, thereby cursed by having a servant for a son was conveniently reclassified as black, and his descendants were hence designated as Africans[2] (Rae, 2018). Accordingly, Europeans, were the blessed white descendants of Noah, while non-whites were the cursed Black descendants of Ham.

For almost 400 years, European slavers loaded approximately 12.5 million African men, women, and children onto slave ships to cross the Atlantic Ocean. About 11 million survived the voyage with about a third going to Brazil (Kachur, 2006). Slave ships were notorious for unimaginably execrable conditions. In 1789 Olaudah Equiano (c. 1745–1797) described his personal experience as a slave on one of those ships in his book *The Interesting Narrative of the Life of Olaudah Equiano, Or Gustavus Vassa, The African Written By Himself* (2005):

> *I was soon put down under the decks, and here I received such a salutation in my nostrils as I had never experienced in my life: so that, with the loathsomeness of the stench, and crying together, I became so sick and low that I was*

*not able to eat, nor had I the least desire to taste anything. I now wished for the last friend, death, to relieve me; but soon, to my grief, two of the white men offered me eatables; and, on my refusing to eat, one of them held me fast by the hands, and laid me across I think the windlass, and tied my feet, while the other flogged me severely.*

(p. 21)

## Colonization

### Colonization as Supported by Slavery

Columbus' inadvertent landing in what came to be known as the "New World" inaugurated systematic European colonization of the Americas. Whatever the original intent that motivated these voyages, Columbus surely opened the door to a new way for European enrichment. Pope Alexander VI's (1431–1503) *Inter Caetera* papal bull of 1493, provided the Spaniards, in whose name Columbus was sailing, with the right to locate, assume dominion over, and take possession of those "lands not possessed by any Christian prince . . . for the honor of God himself and for the spread of the Christian Empire" (quoted in Newcomb, 2008, p. 56). The Atlantic slave trade resulted from, among other things, aggressive European colonization and a consequent labor shortage, itself created by European colonial exploitation of New World lands and resources for great profit; slaves, after all, were unpaid labor.

Importing Africans slaves was not the only means that Europeans used to gain economic power. Indigenous peoples were also captured for slave labor; however, large numbers of them died from overwork and communicable diseases brought from Europe (Lepore, 2018; Wilkerson, 2020). In the New World, however, the slavery of Africans had a self-perpetuating innovation: the children of the millions of enslaved people, transported halfway across the globe with no chance of returning home, automatically became slaves as well.

Shortly after Columbus' ship ran aground on the 6th of December 1492 against the island of Hispaniola (now divided between the Dominican Republic and Haiti), large-scale western European exploration and colonization followed. The Spaniards were the first to build their empire of the Americas using islands such as Cuba, Puerto Rico, and Hispaniola as their commercial and slavery bases. Spanish *conquistadores* managed to take large swathes of North and South America, causing the deaths of a conservatively estimated 8,000,000 indigenous people, primarily through the spread of diseases like smallpox and measles (Ostler, 2015; see also Mann, 2011). At the same time, the Portuguese followed suit temporarily claiming lands in North

America (Canada) and colonizing much of eastern South America, naming it Santa Cruz and Brazil. In the latter part of the 17th century, England belatedly awoke to the potential to colonize North America, when King James issued a charter granting permission to settle in Virginia (1606). Echoing the Roman Catholic papal blessings (Roman Catholic) for the Spanish and Portuguese colonization, the Anglican English crown claimed American land because it was "not possessed by any Christian Prince or People" (Thorpe, 1909). By the 17th century, France and the Netherlands also succeeded in establishing permanent colonies in the Americas. Western Europe portioned the New World as a birthday cake among gluttonous children.

When European colonizers came to the Americas, they conveniently convinced themselves that indigenous peoples were somehow deserving of their fates. Some colonizers felt that God sanctioned the oppression of indigenous people since many died of diseases brought over by the Europeans (Mann, 2011). As was stated by a French settler, "it appears visibly that God wishes that they [Native Americans] yield their place to new peoples." (Crosby, 2015, p. 213). Even when confronted by the fact that colonization was abusive to native and African slaves, some European scholars at the time justified continued inhumanity because the natives were without written language or government; they were therefore uncivilized (Lepore, 2018). Colonizers could find support in texts as old as Aristotle's *Politics* from 350 BCE (Aristotle & Reeve, 1998), in which, foreshadowing the biological determinism of the 20th century, Aristotle avers "From the hour of their birth, some are marked out for subjection, others for rule." Against the acquisition of unheard-of wealth and power, and with philosophical support, it is not surprising that one of Spain's royal historians regarded the differences between indigenous peoples and Europeans as great as those "between apes and men" (Lepore, 2018, p. 25), a trope that unfortunately remained embedded in the racist canon.

## Legal Slavery in the United States and the Moral Dissonance

The English proposed to convert indigenous peoples by, as opposed to baptism, reading them the Bible, an approach the English considered more humane than the forced conversions of the Spanish. The English also brought new philosophies about governing, which moved rulership away from divine right to self-government. For example, English enlightenment philosopher John Locke (1632–1704) stated in 1689 that all are born into a "state of equality, wherein all the power and jurisdiction is reciprocal, no one having more than another," each "equal to the greatest, and subject to no body" (Locke & Goldie, 2016, p. 4). Locke (1689) described all men as coming from a state of perfect freedom independent of the will of others. He wrote: "There being nothing more evident than that creatures of the same species and rank, promiscuously born to all the same advantages of nature, and the use of the same faculties, should also be equal one amongst another without subordination or subjection" (Book 2, p. 106). It was clear that "Slavery is so vile and miserable an estate of man, and so directly opposite to the generous temper and courage of our nation, that is hardly to be conceived that an Englishman, much less a gentleman, should plead for it" (Two Treatises on Government, Book 1, p. 1).

Locke was referring to English peoples as slaves to the divinity of a king, but for exports from Africa, Locke had a very different view, one he expressed in The Fundamental Constitutions of Carolina (1669). There he stated that "Every Freeman of Carolina, shall have absolute power and Authority over his Negro slaves" (Lepore, 2018, p. 55). This contradiction was justified by the innovation of the *concept* of race, in which the imposition of racial categories determined who—or what—constituted a "person." Superficial biological traits like the expression of melanin, the size of the epicanthic fold, and the shape of hair follicles were used to decide significant social questions, such as who belongs and who does not belong to the center of power.

The inception of slavery in America is generally regarded to have occurred August 20, 1619. It was on this day when the privateer *The White Lion* arrived at the British colony of Jamestown, Virginia with twenty African slaves who were traded for food (see Wilkerson, 2020). These twenty men had been seized from a Portuguese slave ship ironically named, given its occupation, after the Catholic saint, *Sao Jao Bautista* "Saint John the Baptist" (McCartney, 2019).

There were dissenting voices. For example, English-born American political activist and philosopher Thomas Paine (1737–1809) stated that since "all men [were] originally equals" the idea of God granting one person and his heirs the right to rule over others was untenable because "nature disapproves it" (1776, p. 7). No matter; in the Americas the slave trade grew exponentially during the 17th and 18th centuries and race became the most widely used demarcation of social status, with whites being superior to all others. European countries conquered and colonized most of North and South America, including the Caribbean islands, which served as slave ports (Lepore, 2018). Expanding colonial territories in the New World and the prospect of owning land[3] rather than working for others, lured many Europeans to settle in the Americas. The migration of Europeans created a huge demand for labor forces for mining coal and farming in the tobacco, cotton, and sugar fields (Blackmon, 2009). Britain sent prisoners to work on the plantations, but these were not sufficient, and to satisfy the tremendous demand for labor, planters purchased slaves (The Abolition Project, 2009; Lepore, 2018). Slavery by Europeans led to generation after generation of (mainly) African peoples being taken from their homelands and enslaved for their and their descendants' lives.

Laws from that time indicate that BIPOC were dehumanized and treated as somehow less than the English. In December of 1662, Virginia's Act XII, written by Virginia's all-white General Assembly, resolved what was seen as a growing problem in the colony. Englishmen were having children with African-descended slaves and there were differences of opinion about whether children of slave mothers were born free or slave. English tradition and law were that a child received his or her status from his or her father. However, free mixed-race children would blur the racial divide forming in Virginia by the 1660s. To resolve this, Act XII declared that "Negro women's children [were] to serve according to the condition of the[ir] mother[s]"—in direct contrast with paternal descent laws that prevailed in England (Shifflett, 1998). This law is significant because it enshrined enslavement as a heritable condition and thus became instrumental in building a racially divided society in which melanin pigment signified bondage. Equally disturbing, it also financially incentivized acts of sexual violence against enslaved women because their children would become white property. This law is then an example of the codification of a white supremacist society evidently envisioned by many colonizers (Jones-Rogers, 2019).

## Capitalism

### Unfettered Markets—Powered by Slaves

Superficially, it may appear counterintuitive that increasing individual and collective liberties would even be related to the regulation of "free" markets. However, when we consider the initial conditions that enabled and even encouraged land theft, dispossession of material wealth, and the kidnapping and subsequent commerce of human lives, it can be appreciated that free markets at the global level are not grounded in freedom as much as in greeddom. In theoretical, true, free markets, there is presumed to be a level playing field in which supply and demand are independent of political and historical context. Under those conditions, the prices for goods and services are presumed to self-regulate. However, because of the historical lack of equity, we continue to pay the price of slavery and colonization—a cost that still disproportionally burdens the BIPOC who have historically been—and continue to be—oppressed in comparison to whites who have enjoyed any number of societal inequities (see Chapter 5).

Since the 18th century, western European nations—and most of their former colonies—have functioned to a greater or lesser degree in a capitalist economy. In this economic system, the means of production are privately owned and owners (including shareholders) are focused on generating personal profits, while those without capital are workers who earn a salary or wage (Zimbalist & Sherman, 2014). Central to capitalism is the recognition of property rights, capital accumulation, wages paid for labor, and a price system based on competitive markets.

English philosopher Herbert Spencer[4] (1820–1903), proposed that the market would be self-regulating; and supply would balance demand (Weinstein, 2019). Spencer ignored historical factors and systemic structures—such as slavery—that create unjust markets in the first place; instead, he chose to imagine a society that would spontaneously arise from self-regulating markets. This, he proposed, would lead to the eventual withering of government since no oversight would be needed (Weinstein, 2019). This idea—ever since Spencer's time, and in its modern incarnation as "Reaganomics"—is still compelling today for those who would see markets unregulated, and government downsized.

Since race-based slavery and colonization were both indispensable drivers of the expansion of wealth for Western European countries and their descendants in the United States, racist policies have long been integral to the expansion of capitalism (Cox, 1948). Marx (1976) presciently wrote in 1847:

> *Slavery is an economic category like any other. . . . Needless to say we are dealing only with direct slavery, with Negro slavery in Surinam, in Brazil, in the Southern States of North America. Direct slavery is just as much the pivot of bourgeois industry as machinery, credits, etc. Without slavery you have no cotton; without cotton you have no modern industry. It is slavery that gave the colonies their value; it is the colonies that created world trade, and it is world trade that is the precondition of large-scale industry. Thus slavery is an economic category of the greatest importance.*
>
> (p. 167)

In many respects, the philosophy of capitalism is inseparable from its historical development, since race and racism played a vital role in developing the formation of the world market enabling the current and inequitable capital accumulation. Kendi (2019) contends that modern capitalism emerged during the 16th century, at the same time as the transatlantic slave trade of African people; hence, he views capitalism and racism as "conjoined twins."

In a 1967 speech delivered at the Southern Christian Leadership Conference, Martin Luther King Jr. said, "Now, when I question the whole society, it means ultimately coming to see that the problem of racism, the problem of economic exploitation, and the problem of war, are all tied together. These are the triple evils that are interrelated" (King, 2018). For Dr. King, the high rates of incarceration of racial minorities, the wage gap, the wealth gap, and the limitation of schooling opportunities, were all indications of an intricate relationship between race and a capitalist system ultimately benefiting owners rather than workers; generally, whites.

Kendi (2019) defines modern capitalism as the freedom to exploit people in order to increase the wealth of the rich,

> *[C]onservative[s] . . . define capitalism as the freedom to exploit people into economic ruin; the freedom to assassinate unions; the freedom to prey on unprotected consumers, workers, and environments; the freedom to value quarterly profits over climate change; the freedom to undermine small businesses and cushion corporations; the freedom from competition; the freedom not to pay taxes; the freedom to heave the tax burden onto the middle and lower classes; the freedom to commodify everything and everyone; the freedom to keep poor people poor and middle-income people struggling to stay middle income, and make rich people richer. The history of capitalism—of world warring, classing, slave trading, enslaving, colonizing, depressing wages, and dispossessing land and labor and resources and rights—bears out the conservative definition of capitalism.*
>
> (p. 162).

He argues that safeguards are required, providing a safety net for all people, preventing monopolies, protecting the environment from damages, regulating big businesses to safeguard consumers; even though these are all seen as anti-capitalist by some conservatives, in an ultra-individualistic, false understanding of the Founders' views of government.[5] Considering the race-based injustices in society, there are profit-motives for essential life-sectors such as education, healthcare, housing, and incarceration. "Capitalism is essentially racist; racism is essentially capitalist," Kendi (2019, p. 163) concludes.

## Castes

As we have seen, to institute and maintain systems that enabled slavery, colonization, and oppressive economic practices, a means of determining social position was required, and race fit the bill. Race became a visible marker of position in the social caste. While Americans have, since the Bill of Rights, affirmed the idea that the United States is a land of equality, Wilkerson (2020) has recently revisited the argument that the United States society can actually be best characterized as a caste system comparable, although not identical, to the caste systems found in India and in pre-"Final Solution" Nazi Germany. She uses the term "caste" because the term "racism" imprecisely captures the immutable social hierarchy into which people are born, and that structures society. She identifies eight "pillars" of caste injustices, shared by all caste societies:

> *{The eight pillars} are the historic origins, the . . . piers beneath the surface of a caste hierarchy. As these tenets took root in the firmament, it did not matter so much whether the assumptions were true, as most were not. It mattered little that they were misperceptions or distortions of convenience, as long as people accepted them and gained a sense of order and means of justification for the cruelties to which they had grown accustomed, inequalities that they took to be laws of nature.*
>
> (Wilkerson, 2020, p. 99)

For her, caste is the bones and race is the skin; therefore, race is ultimately just a visual cue to where one resides in the caste system.

Our current caste system consists of a hierarchical division of our citizenry, that ranks human value in society, determining the standing, respect, access to resources, assumptions of competence and even physical beauty by caste. A gradient of skin color largely determines one's place in the lower part of the caste system[6] (Wilkerson, 2020). The ideas of caste systems have been around for thousands of years and predates the transatlantic slave trade. The caste system we have inherited today intrudes all aspects of life because it controls and restricts where people grow sociologically. Caste is about structure as Wilkerson writes:

> *A caste system is an artificial construction, a fixed and embedded ranking of human value that sets the presumed supremacy of one group against the presumed inferiority of other groups on the basis of ancestry and often immutable traits, traits that would be neutral in the abstract but are ascribed life-and- death meaning in a hierarchy favoring the dominant caste whose forebears designed it. A caste system uses rigid, often arbitrary boundaries to keep the ranked groupings apart, distinct from one another and in their assigned places.*
>
> (p. 17)

In the past, the idea that in the United States race is a synonym for caste was discussed by several other thinkers including abolitionist and Massachusetts senator Charles Sumner (1811–1874). Subsequently, Ashley Montagu (1942) wrote: "When we speak of the race problem in America, what we really mean is the caste system and the problems which that caste system creates in America" (p. 180).

America's castes are, in the first instance, a binary division of whites and non-whites. They have a palpable impact on the lives and the health of people who are relegated to the lower caste and subjected to oppression on a regular basis. Unlike the way Nazi Germany determined who belonged in the "Aryan" race, in the Unites States, it only took "one drop of non-white blood"[7] to be categorized as non-white (Davis, 1991). And that, in turn, leads to an important second implication of the caste system: it is heritable. If you are a member of a particular caste, your offspring will likewise be a member of that caste. And if that caste exists in a state of slavery, your offspring will be slaves. Today, the descendants of slaves, non-white immigrants, and indigenous peoples, are still faced with countless limitations on the freedom to choose the life they lead. Empirically, the Civil Rights Act of 1964 should have concluded the systematic oppression of minorities; however, the deep roots of caste easily overwhelmed legalities.

## Conclusion

Oppression of people has been conducted for millennia but our current social structures are reflective of four relatively recent systematic evils: the transatlantic slave trade, colonization by Western Europe, the current approach to free markets in the context of that history, and the rigidity of our caste system. Economic inequality and discrimination are still prevalent in our market-driven societies, creating and maintaining disparities between races, ethnicities, and socioeconomic groups. Taken together, slavery, colonization, and unfettered free markets have created a legacy of racism that continues to shape society today. This legacy can be seen in the persistent disparities in areas such as education, income, and access to opportunity. Addressing and dismantling these systems of racism requires acknowledging their historical roots and taking active steps to promote equality and justice for marginalized communities.

*As a seven-year old, I did not comprehend why I was being treated differently than my friend, but I had an inkling. Even at that early age, I knew that the pigmentation of my skin was serving me better than his, and that of others, was serving them.*

*Ripping up my award at the* Liceo Salazar y Herrera *was my way of standing up for what I considered a clear injustice. After explaining this to my mother, she gave me a hug and we cried together, me out of anger, and she out of that pride that comes from knowing that her words about fairness and justice were being heard and understood. My time at the* Liceo Salazar y Herrera *ended abruptly when I migrated with my mother to New York. In search of a better future for her family, my mother left her job as a teacher to work a house-keeping job at Super 8 Motel. Surely America, in which opportunities were equally available for all, would be a place where we could thrive!*

## Notes

1 However, as noted by David Baddiel in "Jews Don't Count" (2021), while Jews today (2022) are considered "white," they were certainly considered non-white and less-than human as recently as the 1960s. This highlights the empirically derived fact that the ever-changing compositions of "races" derives from what people need them to be socially and/or politically, rather than from any kind of biological reality.

2 This was a convoluted and genetically impossible story as all three were biologically Noah's sons with similar genotypes and phenotypes. Despite its inconsistencies, as obvious then as later, it was a validation for skin color and ancestry as the demarcation between slavery and freedom.

3 In European societies since well before the Middle Ages, in contrast to some other cultures, land ownership was seen as a key to wealth and was ultimately tied to higher castes (nobility). As shown in Chapter 3, land ownership became—and remains—a key to wealth in the New World as well.

4 Spencer also coined the phrase "survival of the fittest," a reference to the idea that economic success depended upon the supply/demand equation; that too much supply with no demand was ultimately a recipe for business failure. An early proponent of Darwinian evolution, he and many others subsequently—and incorrectly—applied the expression to organisms as well (see Chapter 7).

5 Termed "free-market fundamentalism," the idea has been popularized—particularly during the Reagan Presidency—that the Founder's view of government was, in effect, less is more; and that in a perfect capitalist society, markets would self-regulate, such that many of the evils of capitalism—huge wage disparities; monopolies; worker abuse—would disappear. This idea, despite continued traction among constitutional "originalists " in the Republican Party, has been convincingly discredited by authors N. Oreskes and E.M. Conway (2023), who demonstrated that far from free-market fundamentalism, America's Founders recognized that the free-market economy of the United States has required and will continue to require guardrails: federally mandated regulations on the market.

6 That Americans—even Americans within the white caste—feel this way is revealed by a recent widespread attraction to genealogy. There is a strong

interest in who we *were*—Italian, French, English—rather than who we *are* (American). And within the white community, is it not true that people whose relatives go back to the Founders are in some way imbued with a higher status than those whose relatives arrived with the great immigration events of the early 20th century? Ironically by this metric, because persons of color have been here since the early 1600s, a Black person with ancestors traceable back to Thomas Jefferson (or even earlier) should occupy a much higher caste than most of the whites that inhabit the United States. But color as a caste determinant in the United States trumps everything else.

7 An infamous but widely held sentiment, immortalized Jerome Kern's and Oscar Hammerstein's 1927 musical "Showboat," in which Sheriff Vallon, looking to shut down a show because of miscegenation between two supporting actors, says, "one drop of n—r blood makes you a n—r in these parts."

## References Cited

The Abolition Project. (2009). *The Arrival of European Traders*. Retrieved August 5, 2020, from http://abolition.e2bn.org/slavery_42.html

Aristotle, & Reeve, C.D. (1998). *Politics*. Cambridge: Hackett Publishing Company.

Blackmon, D.A. (2009). *Slavery by Another Name: The Re-Enslavement of Black Americans from the Civil War to World War II*. New York: Anchor Books.

Boddy-Evans, A. (2020, July 25). *A Short History of the African Slave Trade*. www.thoughtco.com/african-slavery-101-44535

Cox, O.C. (1948). *Race, Caste and Class: A Study in Social Dynamics*. New York: Doubleday.

Crosby, A.W. (2015). *Ecological Imperialism: The Biological Expansion of Europe, 900–1900*. Cambridge: Cambridge University Press.

Davis, F.J. (1991). *Who Is Black? One Nation's Definition*. University Park: Penns State University Press.

Equiano, O. (1789/2005). *The Interesting Narrative of the Life of Olaudah Equiano, Or Gustavus Vassa, the African Written By Himself*. The Project Gutenberg EBook. www.gutenberg.org/files/15399/15399-h/15399-h.htm

Feagin, J.R. (2006). *Systemic Racism*. New York: Routledge.

Fredrickson, G. (2002). *Racism: A Short History*. Princeton; Oxford: Princeton University Press. https://doi.org/10.2307/j.ctvc779fw

Ince, I.J. (2020, June 11). *DC Pastor on the Role of Diversity in Places of Worship*. Retrieved July 02, 2020, from www.npr.org/transcripts/874568656

Jones-Rogers, S.E. (2019). *They Were Her Property: White Women as Slave Owners in the American South*. New Haven: Yale University Press.
Joshel, S. (2019, October 15). *Roman Slavery and the Question of Race*. Retrieved August 4, 2020, from www.blackpast.org/global-african-history/perspectives-global-african-history/roman-slavery-and-question-race/
Kachur, M. (2006). *Slavery in the Americas: The Slave Trade*. New York: Infobase.
Kendi, I.X. (2017). *Stamped from the Beginning: The Definitive History of Racist Ideas in America*. New York: Bold Type Books.
Kendi, I.X. (2019). *How to Be an Antiracist*, 1st ed. New York: One World.
King, M.L. (1967/2018, March 28). *Where Do We Go From Here?* Address Delivered at the Eleventh Annual SCLC Convention. Retrieved August 7, 2020, from https://kinginstitute.stanford.edu/king-papers/documents/where-do-we-go-here-address-delivered-eleventh-annual-sclc-convention
Lepore, J. (2018). *These Truths: A History of the United States*. New York: W.W. Norton & Company.
Locke, J. (1689). *Two Treatises on Government*. London: Printed for R. Butler, etc., 1821; Retrieved March 31, 2019, from www.yorku.ca/comninel/courses/3025pdf/Locke.pdf
Locke, J., & Goldie, M. (2016). *Second Treatise of Government and a Letter Concerning Toleration*. Oxford: Oxford University Press.
Mann, C. (2011). *1493: Uncovering the New World Columbus Created*. New York: Knopf.
Marx, K. (1976). The Poverty of Philosophy. In *Marx–Engels Collected Works*, Vol. 6. New York: International Publishers.
McCartney, M. (2019, October 8). Virginia's First Africans. *Encyclopedia Virginia*. www.EncyclopediaVirginia.org/Virginia_s_First_Africans
Montagu, A. (1942). *Man's Most Dangerous Myth. The Fallacy of Race*. Walnut Creek, CA: Altamira Press.
Newcomb, S.T. (2008). *Pagans in the Promised Land: Decoding the Doctrine of Christian Discovery*. Golden, CO: Fulcrum Publishing.
Oreskes, N., & Conway, E.M. (2023). *The Big Myth: How American Business Taught us to Loathe Government and Love the Free Market*, p. 576. New York: Bloomsbury Publishing.
Ostler, J. (2015). Genocide and American Indian History. *Oxford Research Encyclopedia of American History*. https://doi.org/10.1093/acrefore/9780199329175.013.3
Paine, T. (1776, February 14). *Common Sense Addressed to the Inhabitants of America*. Retrieved March 17, 2019, from www.learner.org/workshops/primarysources/revolution/docs/Common_Sense.pdf
Rae, N. (2018). *The Great Stain: Witnessing American Slavery*. New York: The Overlook Press.

Shifflett, C. (1998). *Virtual Jamestown—Laws on Slavery*. Retrieved August 10, 2020, from www.virtualjamestown.org/laws1.html

Thorpe, F.N. (1909). *The Federal and State Constitutions, Colonial Charters, and Other Organic Laws of the State, Territories, and Colonies now or Heretofore Forming the United States of America*. Washington: Government Printing Office.

Weinstein, D. (2019, Fall). Herbert Spencer. In E. N. Zalta (Ed.), *The Stanford Encyclopedia of Philosophy*. https://plato.stanford.edu/archives/fall2019/entries/spencer/

Wilkerson, I. (2020). *Caste: The Origins of Our Discontents*. New York: Random House.

Zimbalist, A.S., & Sherman, H.J. (2014). *Comparing Economic Systems: A Political Economic Approach*. New York: Elsevier Science.

Zurara, G.E. (1453/2010). *The Chronicle of the Discovery and Conquest of Guinea*. Cambridge: Cambridge University Press.

# 9

# Oppression in the United States

*My two sons are very different from each other. That means that I have very different conversations with each of them. Isaac, who was born with Down syndrome, really prefers to talk about Weird Al Yankovic, or the medical procedures portrayed in the television drama "House." But Nicolas has a weak stomach for blood. He lacks that appreciation of human anatomy that his older brother Isaac has. Nicolas is much more mathematical and methodical in his thinking. He can quickly do impressive mental math and engage in discussing global events. I am responsible for ensuring that both of these boys have access to the kinds of resources I often lacked as a child. I am committed to the idea that they will be part of a world that values humanity more than our current society.*

*Nicolas and I often find ourselves speaking about those "big" ideas in life. Some of these discussions can be quite challenging for me, as I want him to recognize that, like all humans, I cannot be completely unbiased and that my words are not the ultimate source of truth. One of the most difficult conversations I have had with Nico was the one where I had to point out to him that as a Latinx in the United States, some people will hate him before they even meet him. The forty-fifth President's rhetoric about Mexicans and Mexican-Americans, who are the majority of Latinx people in the United States, has emboldened hatred toward all Latinx people. The former President's words carry weight and so, on August 19, 2015, two white men who said they were inspired by President Trump beat and urinated on a homeless Latinx man. The President responded by saying that his supporters were "passionate" (Moyer, 2015). More recently the results of his hate-filled speech and ideology came on August 3rd, 2019, when Patrick Crusius drove 650 miles to use an AK-47 assault rifle to murder 23 people and injure 23 others with the intent of "shooting as many Mexicans as*

DOI: 10.4324/9781003409144-13

*possible" (Hutchinson & Mallin, 2020). Subsequently, Crusius' internet screed was found to mirror comments Trump had made in the past, including references to illegal immigration as an "invasion" and telling a group of BIPOC congresswomen (all American citizens) to "go back to [their] home countries" (King, 2019).*

*Words failed me as I told my son that some people want to kill us solely based on our ethnic heritage. Nico did not choose to be born a Latinx in the United States. At the time that Patrick Crusius began his shooting spree, my son was fourteen years old. It was difficult expressing to him that neither his abilities nor his talents, his efforts nor his dedication, his potential nor his qualities would matter, his presumed background and appearance alone would have made him a target for Crusius. What historical events made this conversation necessary? Why does this happen in a country that is supposed to be exceptional and unique, representing the best hopes of the world for humanity? What is holding us back from the quest set forth in the Inaugural Address of President John F. Kennedy in 1961, who challenged Americans to "explore what problems unite us instead of belaboring those problems which divide us." In this chapter we discuss oppression as it is expressed in the United States.*

*Growing up with two languages meant that there were some concepts that made more sense to me exclusively in Spanish or English—depending on the context. My mother taught me that there are two kinds of people who perpetrate evil, the* **descarados**, *and the* **solapados**. *According to her, at times all of us are at risk of behaving as* descarados *or* solapados. *Both are to be avoided since the results of both are the maltreatment of others, but these two terms differ from each other based on the approach to the wrongdoing they represent. In English* descarado *can be translated as shameless, brazen, or blatant. For my mother,* descarados *are people who wrong others, without hiding their intentions and motives, and comfortable with the knowledge of what they are doing. An obvious example of* descarado *behavior was the legally-sanctioned enslavement of Africans in the Americas. By contrast, the English translation of* solapado, *"underhanded," fails to capture the full meaning of the word. This was the term my mother used when someone, also in complete awareness, used malicious intent to harm others, while appearing entirely innocent or even beneficent. We shall see that oppression in the Americas has been and remains both* descarado *and* solapado.

## The Time of "*Descarado*"—When Slavery was American Law

Despite a backlash of revisionism, epitomized by the "Lost Cause" of the Confederacy, that seems to have beset the United States since the summer of 2020, it is our view that only a cold, hard look at our racist (e.g., Lepore,

2018), and imperialist (e.g., Immerwahr, 2019), past will allow us to make properly informed decisions going forward. Accordingly, these subjects need to be part of the conversation as we understand the way that science has been used to justify racist and imperialist policies.

## In the Beginning . . .

From its foundation, the United States was a country with racial inequalities. As they do today, ideas of race and power varied in the colonies and early days of the United States. On one end of the spectrum, Lepore (2018) describes radicals like pamphleteer Benjamin Lay who viewed slavery as an absolute evil. Somewhere in the middle, Benjamin Franklin owned slaves but promised their freedom after his death. And on the opposite end of the spectrum, were the southern slave-owners whose very economy, and thus, livelihood, depended upon slavery. They and could not see their economic future without access to unpaid labor (Lepore, 2018). The concept of race was felicitous for this last group, because although the Declaration of Independence professed that the "self-evident" truth of human equality, the protective ægis of science biologically justified the existence of slaves, who, being less than fully human, were less than fully equal. Herein lies the great ambiguity—an abolitionist such as William Lloyd Garrison might have termed it an "hypocrisy" of the US social contract—in which reason could embrace both equality for all "men" along with the institution of slavery.

Legally-sanctioned slavery was part of the United States from its foundation until 1865 and during that time it had staunch defenders. In explicit language that still shocks, Alexander Stevens,[1] Vice President of the Confederate States of America, said of the newly formed Confederacy:

> *Its foundations are laid, its cornerstone rests upon the great truth that the negro is not equal to the white man; that slavery subordination to the superior race is his natural and normal condition. This, our new government, is the first, in the history of the world, based upon this physical, philosophical, and moral truth. . . . With us, all of the white race, however high or low, rich or poor, are equal in the eye of the law. Not so with the negro. Subordination is his place. He, by nature, or by the curse against Canaan, is fitted for that condition which he occupies in our system.*
>
> (as quoted in Lepore, 2018, p. 290)

Here, then, is the Vice President of the "Confederate States of America" justifying by biology ("by nature") and/or by the Christian religion ("by the curse of Canaan") Black enslavement.

In spite of Stevens' (and his confederates) strongly held views, the surrender of the Confederacy on April 9, 1865, operationally ratified, as it were, the Emancipation Proclamation of 1863, freeing all slaves in the United States. On December 6, 1865, the House of Representatives legally ratified the 13th Amendment which abolished slavery in the United States. *Descarado* slavery had existed on the American continent for more than 200 years, and with this new law, some 4 million African American men, women, and children were pronounced free.

Of course, with a 200-year history of slavery, the former slave-holding states' economies in ruin, and many peoples' mindsets still firmly imbued with the biological inferiority of non-whites, simply pronouncing people as "free" really didn't magically solve their problems. Most of those who were freed in 1865 had no home, no money, and no way to earn an income. Promises were made to the newly freed people, including Special Field Orders No. 15, a wartime order proclaimed by Union General William Tecumseh Sherman (1820–1891) on January 16, 1865, which provided for the settlement of 40,000 freed slaves and black refugees on land expropriated from white landowners in South Carolina, Georgia, and Florida (Reidy, 2019). Sherman later ordered the army to lend mules for the agrarian reform endeavor, in an effort famously known as "Forty acres and a mule." Also, to address the needs of the newly freed slaves, the US government created the Bureau of Refugees, Freedmen, and Abandoned Lands. Commonly known as "the Freedmen's Bureau," it provided food, housing, and medical assistance to thousands of freed slaves.

The assassination of Abraham Lincoln on April 15, 1865, dramatically changed the fortunes of people of color in the United States. The prewar South Carolina plantation owners (and their representatives in Congress) who had fled their properties earlier in the Civil War now demanded the restoration of their estates. Lincoln's former Vice President (now President) Democrat (Tennessee) Andrew Johnson was unsympathetic to Lincoln's ideas for the reconstruction of the former Confederacy. In October of 1865, he sent General Oliver Otis Howard (1830–1909), Commissioner of the Freedmen's Bureau, to coastal South Carolina to ensure that the former slave owners received their lands back. Howard proposed what he saw as a kind of compromise: the freed people could accept labor contracts with their former owners instead of the land they had been promised. Of course, this frustrated the possibility of freed people owning land, as well as indebting them, for their livelihoods, to their former owners. This effectively restored the pre-war power and opportunity inequities that were at the root of slavery (Reidy, 2019).[2]

For approximately 10 years, the (generally Republican) forces of reconstruction warred against the (generally southern Democratic) forces resisting Black equality. Ultimately, the death knell for Republican-driven,

Lincoln-inspired, but highly contentious Reconstruction sounded in 1876, when a truly unholy compromise was reached: Democrats would support Republican Rutherford B. Hayes' bid for the Presidency, on the condition that federal troops that had been stationed in the south since the end of the Civil War to ensure the security of the Reconstruction and the protection of former slaves, be withdrawn. And with their departure, Reconstruction became what Andrew Johnson had preferred to call "restoration" (Lepore, 2018, p. 318), and the "Jim Crow" period began for blacks in the South. Night had truly descended upon race relations in the United States.

After 1876, the courts and Congress chipped away at the Reconstruction Amendments' (the 14th and 15th) promises to Blacks—suffrage, equal protection, and due process before the law. The most iconic legislation, capping this steady erosion of rights, was the notorious Plessy vs. Ferguson Supreme Court decision of 1896. In 1890, Louisiana law stipulated that all passenger railways had to have ways of "providing for separate railway carriages for the white and colored races" (Bancroft Davis, 1896, p. 537). Homer Adolph Plessy was of mixed race and took a vacant seat in a whites-only car. After refusing to leave the car at the conductor's insistence, he was arrested and jailed. Subsequently convicted by a New Orleans court of violating the 1890 law, Plessy filed a petition against the presiding judge, Hon. John H. Ferguson, claiming that the law violated the Equal Protection Clause of the 14th Amendment (Vann Woodward, 1964). The case went to the United States Supreme Court, and Plessy lost. The Plessy vs. Ferguson opinion embraced a "separate but equal" doctrine: races should be separate in all respects including marriage, but conditions, that is, the living accommodations (schools, housing, jobs, churches, restrooms, hotels, restaurants, hospitals, cemeteries, etc.) accorded to each were to be equal. In practice this promise of equality of conditions was never remotely fulfilled, and thus the Plessy vs. Ferguson decision was surely *solapado*: while seeming to be equitable on paper, in practice it signaled to whites and non-whites alike that despite the Constitution's guarantees, the fundamental rights of non-whites would not be protected. The letter of the law, and even the Constitution, was insufficient since it could not be used to defend the rights of BIPOC. Laws were enacted that were aimed to oppress non-whites and, in the process, created immense suffering.

Along with Africans, Indigenous peoples, also seen as effectively sub-human (that is, along the venerable *scala naturæ*—see Chapter 10—more bestial than fully human), were thus not exempt from overt *descarado*, with the goal of removing them from their lands and their effective separation from white society and institutions. This quasi-genocidal approach had been pioneered by the Spanish in the 16th century New World and Dutch in 17th century Indonesia, and thus it was not without historical precedent that as late as

November 29, 1864, a regiment of Colorado volunteer cavalrymen mass murdered an encampment of Cheyenne and Arapahos at Sand Creek, Colorado, killing about 200 people, mostly women and children. This came to be known as the Sand Creek Massacre. When the troops arrived, a Cheyenne chief raised the Stars and Stripes above his lodge, while others in the village waved white flags signifying that they were not going to battle the seven hundred blue-clad cavalrymen. Ignoring their plea, the American troops, led by a pistol-packing minister (John Chivington; 1821–1894), opened fire with carbines and cannon, and before leaving, burned the village and mutilated the dead, carrying off body parts as trophies (Horwitz, 2014). Thanks to Capt. Silas Soule (1838–1865), who served in this regiment and witnessed the attack but was unwilling to shoot the Cheyenne, we know about the evils perpetrated. Soule described hundreds of women and children begging on their knees for mercy but murdered by the white men. Ultimately Soule himself was murdered for shedding light on this atrocity. Similarly *descarado*, at Wounded Knee, South Dakota, on December 29, 1890, the US Army's 7th Cavalry surrounded a band of Ghost Dancers under Big Foot (1826–1890), a Lakota Sioux chief, and demanded they surrender their weapons. As that was happening, a fight broke out between an American Sioux and a US soldier and a shot was fired, although which side fired first is still disputed. A brutal massacre followed, in which an estimated 150 Indians were killed, nearly half of them women and children (Utley, 1963). These are but two examples of ways in which Native Americans have been overtly abused after the dispossession of their lands; truly *descarado* oppression.

Importantly, the legal abolishment of slavery in 1865 did not end the overt oppression of the *descarado* time. There is a plethora of examples in which white people overtly—sometimes even ostentatiously, oppressed BIPOC after 1865. For example, on November 10, 1898, the Wilmington (NC) Coup took place. This is the only instance in the history of the United States in which a legitimately elected government was overthrown by *coup d'etat*. Wilmington had a sizable Black middle class and a multiracial government that angered white supremacists, who stockpiled weapons and promoted false reports of Black violence in the lead-up to the 1898 municipal election (Zucchino, 2020). Newspapers reports incited white citizens' fears but although the federal government was alerted of possible confrontation, the Black citizens of Wilmington were left to fend for themselves. On Election Day, the white supremacists took the government by force, making all Black politicians resign and instating a "White Declaration of Independence" (Zucchino, 2020). Subsequently, a riot ensued and over 2,000 Black citizens were driven out of the city and 60 were killed. These actions sent a clear and lasting message that Blacks were unwelcome and resulted in the city's Black population going from 56% in 1898 to 18% in 2018 (Zucchino, 2020). Clearly *descarado* behavior.

Equally blatantly, and in violation of even the *solapado* "Separate but equal" doctrine of the Plessy vs. Ferguson decision, Southern states enacted laws that denied blacks the rights to testify against whites, to serve on juries or in state militias, to vote, or to start a job without the approval of the previous employer (Forde & Bowman, 2017). In fact, whites in power found innumerable ways to circumscribe the lives of African Americans, including contract laws that penalized Blacks attempting to leave a job before an advance had been worked off, and so-called "Pig Laws" which harshly and disproportionally penalized BIPOC for crimes such as stealing a farm animal. Additionally, vagrancy statutes made it a crime to be unemployed and many misdemeanors or trivial offenses were treated as felonies, with excessive sentences and fines. These laws only applied to non-whites, forcing many Blacks to work for little or no pay, sometimes for the same plantation owners that at one point had owned slaves. In some states Black men who were arrested for not having a job could be hired out to plantation owners creating a huge market for convict leasing (Mancini, 1996). These "free" people, being prisoners, naturally worked for little or no pay, under even more abhorrent working conditions than those to which they were exposed to during legal slavery. Plantation owners leasing the black prisoners justified their ill-treatment because the owners had no long-term interest in the convicts' well-being.

### Into the 20th Century: Continued, Descarado Oppression of BIPOC

Laws reflect the political climate of their time. By the turn of the 20th century every state in the South had mandated racial segregation by law. And the Supreme Court validated these laws with the Plessy vs. Ferguson (separate but equal) decision, until 1954 with Brown v. the Board of Education.

President Calvin Coolidge (1872–1933) signed the Johnson-Reed Immigration Act on May 24, 1924. This act established the harshest immigration restrictions in US history, banning Asians, who were already subject to exclusion laws, from naturalization entirely, and established immigration quotas that radically reduced the numbers of permitted Southern and Eastern European migrants, who were deemed less "white" than other Europeans (Jacobson, 1998).

In the 1930s unemployment rose to record levels during the Great Depression, and Mexican migrants and Mexican-Americans were unjustly blamed for taking jobs from US citizens, many of whom, paradoxically, were living on public welfare (Minian, 2018). In response, the US government's immigration officials started deportation campaigns to remove unauthorized migrants, and for those people who could not be deported because of legal residency or citizenship, there was pressure to leave "voluntarily." With the support of the Mexican government, US officials sponsored trains to return ethnic Mexicans to the border (Minian, 2018). The number of repatriated Mexicans has been estimated from 350,000 to as high as 2 million, out of which 60% are believed to have been American citizens and children (Minian, 2018). Ironically, only a very few years after the final episode of repatriation in 1939–40, US officials were desperate to bring Mexican workers back to replace American citizens who had gone to fight World War II.

Post-war, the notorious murder of Emmett Till (1941–1955)[3] on August 28, 1955, showed that in the world of Jim Crow America, very little had changed. While on vacation from Chicago to visit family in Money, Mississippi, Emmett was shopping at a store owned by a white couple, Roy (1931–1994) and Carolyn Bryant (1934–), and someone said Till whistled at Mrs. Bryant. At approximately 2:30 AM on August 28, 1955, Roy Bryant, and his half-brother J.W. Milam (1919–1980) kidnapped Till. Subsequently, the two white men beat the black teenager brutally, dragged him to the bank of the Tallahatchie River, shot him in the head, tied him with barbed wire to a large metal fan and shoved his body into the water (FBI, 2016). In September 1955, an all-white jury found Bryant and Milam not guilty of Till's kidnapping and murder. Protected by law against double jeopardy because they could not be prosecuted for the same crime twice (American Experience, 2020), Bryant and Milam publicly admitted in a 1956 interview with Look magazine that they had killed Till. Milam is directly quoted in that publication (Huie, 1956):

> *Well, what else could we do? He was hopeless. I'm no bully; I never hurt a nigger in my life. I like niggers—in their place—I know how to work 'em. But I just decided it was time a few people got put on notice. As long as I live and can do anything about it, niggers are gonna stay in their place. Niggers ain't gonna vote where I live. If they did, they'd control the government. They ain't gonna go to school with my kids. And when a nigger gets close to mentioning sex with a white woman, he's tired o' livin'. I'm likely to kill him. Me and my folks fought for this country, and we got some rights. I stood there in that shed and listened to that nigger throw that poison at me, and I just made up my mind. 'Chicago boy,' I said, 'I'm tired of 'em sending your kind down here to stir up trouble. Goddam you, I'm going to make an example of you—just so everybody can know how me and my folks stand."*
>
> (Look Magazine, January 24, 1956, pp. 46–50)

Although the lynching of Blacks in the South was a familiar fact of life there,[4] when Emmett's mother displayed his lifeless, mutilated body to be photographed without make-up in an open casket for the world to see, it was the catalyst that set the civil rights movement of the late 50s and early 60s in motion (FBI, 2016).

On the other side of same racist coin was the difficulty, in the Jim Crow South, of getting a fair trial. The trials of the "Scottsboro Boys" exemplifies this inequity. In 1931 nine Black teenagers were accused of raping two white women on a train in Alabama. Concerns stemming from their conviction included all-white jurors, lack of evidence, lynch mobs before the suspects had been indicted, rushed trials, cross-burning, and anti-Semitic remarks directed at the defense lawyer (Scottsboro: An American Tragedy, 2000). In the end, in the face of multiple rounds of trials, recanted testimony, and clear evidence that the innocence of the Scottsboro Boys, the State of Alabama dropped the charges against four (but after time had been served). All were pardoned in 2013, with then Governor R.J. Bentley stating,

> *While we could not take back what happened to the Scottsboro Boys 80 years ago, we found a way to make it right moving forward. The pardons granted to the Scottsboro Boys today are long overdue.*
>
> (as quoted in Lyman, 2013)

The kind of overt hatred displayed by individuals and commonly backed by federal and/or state authority, can show up in instances when oppressed people are making headway with their struggles. Such is the case with the 1969 attacks on the Black Panther Party's breakfast program. The Black Panther Party was a political organization founded in 1966 by Huey Newton (1942–1989) and Bobby Seale (1936–) to challenge police brutality against BIPOC. As part of their work, the Black Panther's Free Breakfast for Children Program fed tens of thousands of hungry kids in cities across the Unites States (Collier, 2015). The program was proving successful and the FBI and local police feared the Panthers were growing in popularity and worked to dismantle the program. In 1969 in Baltimore, police raided the breakfast program with guns drawn (Bloom & Marin, 2016). That same year, Chicago police broke into the church that hosted the breakfast and smashed and urinated on the food; while in Harlem, police spread false information that the food was poisoned (Heynen, 2009). Ironically, the USDA used the Breakfast Program as a model for a federal school breakfast program (Heynen, 2009).

What was one to think when, two days after the 1963 "I have a Dream" speech, FBI Deputy Director William Sullivan characterized Martin Luther King (a Christian minister and a relentless proponent of non-violence) as "the most dangerous Negro in this nation from the standpoint of communism, the Negro, and national security" (Risen, 2023)? As noted by Silverman (2018),

> Sullivan makes a conscious decision to identify King as "the most dangerous Negro." The interjection of race was not necessary as Sullivan

could have classified King as a most dangerous man, person, or leader. Sullivan inserts "Negro" to connect King's racial background to "dangerous," an adjective that is regularly used to define criminals.

## The Time of "Solapado"—Oppression Under-Disguise

Although post-Civil-Rights Act (1964) policies and laws no longer overtly express the intent to oppress non-whites, there are still systemic forces and official policies which benefit whites over BIPOC—this is under-disguise oppression: *solapado*. For example, as a response to the drug epidemic of the early 1980s Presidents Reagan (1911–2004) and H.W. Bush (1924–2018), from 1981 to 1991, militarized and criminalized the anti-drug approach. The drug enforcement agency's budget to fight criminality and drug use went from $220 billion in 1981 to $875 billion by 1991 (DEA, 2020). Furthermore, the 1986 Anti-Drug Abuse Act which carried mandatory minimum sentences was much harsher for the distribution of crack cocaine which was associated with Blacks than powder cocaine which was associated with whites (Sterling, 2014). State and local spending on prisons and jails has increased at triple the rate of funding for public education in the last three decades, since 1990, spending on postsecondary education has been largely flat while spending on correctional institutions has increased 89% (U.S. Department of Education, 2016). In the United States, Black youth are more than five times as likely to be detained or committed compared to white youth, and in at least six states, BIPOC youth are at least 10 times as likely to be held in placement as white youth (Nellis, 2017). Our current legal system is no longer explicitly racist, but it is *solapado*, showing that American justice is certainly not blind.

Perhaps one of the most egregious examples of *solapado* in recent times comes in the form of voter suppression during and after the 2020 elections (Kroll, 2020). During the global COVID-19 pandemic, the Trump reelection campaign and the Republican National Committee spent over $20 million on voting-related lawsuits in 2020 in more than a dozen states, including the battlegrounds of Colorado, Minnesota, Michigan, Wisconsin, Pennsylvania, and Florida (Kroll, 2020). President Trump attempted to suppress voters using several methods, including the intimidation of voters (Villa, 2020), spreading misinformation about voter fraud (Ortega & Bronstein, 2020), and perhaps most notoriously, by using his political and executive power to limit funding for the United States Postal Service to make it more difficult to deliver mail-in ballots (Levine, 2020). All of these moves were claimed to address voter fraud which even by the report from the conservative Heritage Foundation only happens in extremely rare cases, since out of 250 million votes cast by mail over the past 20 years, less than 1,300 cases (0.00006%) were fraudulent (Samalis-Aldrich & Von Spakovsky, 2020). Several groups filed lawsuits against Trump and Postmaster General Louis DeJoy, alleging that the two violated voters' rights through a series of controversial changes at the US Postal Service that triggered widespread delays (Durkee, 2020). While these tactics were justified by the Trump administration as cost-saving measures for the US Postal Service, they were in fact intended to impair the ability of racially minoritized people to cast their votes (presumably for Joe Biden, the Democratic candidate), which benefitted Trump and his allies (McCaskill, 2020; Cobb, 2018). Additionally, a way to legally suppress votes has been the Tennessee law which punishes protesters, in particular Black Lives Matter protesters, by taking away their right to vote (Robinson, 2020). Ultimately, these voter-suppression efforts had a real impact, as non-white voters are seven times more likely than white voters to wait in line for more than an hour to vote (Pettigrew, 2017).

Another *solapado* effort by the Trump administration was the manipulation of the US Census. Ideally, the census counts every person residing in the United States—citizens, noncitizen legal residents, and unauthorized residents. Census results are particularly important in determining the number of seats in the House of Representatives, and for population-based federal funding formulas.

In 2017, a question about citizenship was added to the census by Trump appointee Wilbur Ross, but in 2019, the Supreme Court decided that the reasons given for the additional question were not sufficient (Liptak, 2019). In response, President Trump wrote on Twitter that he had "asked the lawyers if they can delay the census, no matter how long, until the United States Supreme Court is given additional information from which it can make a final and decisive decision on this very critical matter" (Trump, 2019). The concern was that the citizenship question would stoke fear among unauthorized immigrants and their family members, scaring off millions of mostly Latinx residents from responding, resulting in undercounts in states where those residents are concentrated (Prokop, 2019). After the legal defeat of the citizenship question, the Trump administration found a "work-around" to undermine the accuracy of the 2020 census: the count was ended earlier than planned, during the COVID-19 pandemic, a move that likely resulted in undercounting BIPOC and others in hard-to-reach communities (Wolf, 2020).

With an eye toward the 2024 election, parts of the Republican party have redoubled efforts in voter suppression, largely under the guise of removing election "fraud," a non-issue that was effectively invented before the 2020 election by the Trump administration to ensure that if he lost, he could keep his power (Feldman et al., 2020). It since morphed into a slew of legal challenges and GOP-led legislation involving narrowing voting opportunities in states like Georgia, Texas, and Alabama—*solapado* because although this has been alleged to preserve the "integrity" of the electoral process, it is actually about ensuring that people who would likely vote Democratic—and Blacks & Hispanics constituting by far the largest proportion of this group—will be unable to vote easily (Pilkington, 2021).

The results of *descarado* and *solapado* atrocities remain with us today in the persistence of societal inequities. Although slavery has been outlawed and direct colonization by European states is no longer politically viable, there are deep-seated remnants of these oppressive systems in ways that ensure the continued oppression of BIPOC. Today, the descendants of slaves, non-white immigrants, and indigenous peoples, are still faced with countless limitations of the freedom to choose the life they lead. Empirically, if the legal end of slavery truly concluded the systematic oppression of BIPOC then there ought to be equitable representation in all areas of society. The America of today cannot exist without the history of appropriation of lands and the enslavement of racially minoritized peoples. For instance, the Black labor force played a fundamental part in the physical building of America; a labor force that was enslaved due to a system that saw non-whites as inferior, less than equal and deemed less than fully human.

## Racism and American Politics: "This is a Country for White Men"

Racism and politics in the United States have had a synergistic relationship that has often sought ever-shifting validation. President Andrew Johnson, who inherited Abraham Lincoln's Reconstruction, stated in 1866, "This is a country for white men, and by God, as long as I am president, it shall be a government for white men" (as quoted in Trefousse, 1997, p. 236).

At least 12 US Presidents enslaved Black men, women, and children at some point in their lives (White House Historical Association, 2020) and many more presidents normalized "white supremacy" with their words and actions (Petrella & Hasan Loggins, 2017). From the very beginning of our democracy, the first president, George Washington, kept Black slaves and rotated them between Philadelphia (the capital at that time) and his plantation in Virginia every six months, to ensure that these black people could not sue for freedom after more than six months in the state as was the law in Pennsylvania. Further cementing this legacy, on the 31st of May 1779, Washington wrote a descriptive letter to Major General John Sullivan asking him to devastate Native Americans:

> *The expedition you are appointed to command is to be directed against the hostile tribes of the six nations of Indians, with their associates and adherents. The immediate objects are the total destruction and devastation of their settlements and the capture of as many prisoners of every age and sex as possible. It will be essential to ruin their crops now in the ground and prevent their planting more.*
> (p. 716)

Similarly, many other presidents have voiced racists views, including Andrew Jackson, Thomas Jefferson, Abraham Lincoln, James Garfield, Andrew Johnson, Ulysses S. Grant, Rutherford B. Hayes, Grover Cleveland, Herbert Hoover, and Calvin Coolidge (Petrella & Hasan Loggins, 2017). Sadly, other presidents have voiced their approval of white supremacists, such as the case with Woodrow Wilson who described the KKK as "great" and "veritable," (Petrella & Hasan Loggins, 2017), and infamously admired DW Griffith's racist 1915 caricature of the Reconstruction, "Birth of a Nation."

However, for all the dismantling of Reconstruction, and the imposition of "Jim Crow" conditions, the election of President Barak Obama was seen by most as evidence that the United States had finally shed its racist past and had entered a post-racial stage. Yet, the subsequent election of Donald Trump and the shocking events and policies that ensued, however, interpreted by many commentators as a backlash against the election of a Black President, relentlessly ratified white supremacy with brutal and ongoing consequences (Petrella & Hasan Loggins, 2017).

Trump's presidency produced shocking examples of racially-motivated political attacks. These include Muslim immigration bans, the canceling of international student visas, the gutting of the Voting Rights Act, the separation of children from their parents at the border, the backing of the Charlottesville supporters, the push for a wall on the border with Mexico[5] (see Arce, 2019), and the racist dismissal of COVID-19 as the "kung flu" (Lee, 2020). Trump's re-election campaign consistently repeated the racist and ugly specter of black and brown people invading the suburbs and threatening white women (Glanton, 2020). Unforgettably, after the white nationalist gathering in Charlottesville in 2017, Trump said of the neo-Nazis and the protestors that there were "very fine people on both sides" (Coaston, 2019).

This litany of unsupportable racial policies and statements would not be complete without a brief reflection on the final two weeks of President Trump's Presidency. On January 6, 2021, a right-wing, overtly white-supremacist mob, agitated by months of baseless claims that the Presidential election was "stolen" from Trump, and inflamed at a rally by the President himself, his son (Donald Jr.), and his personal lawyer (Rudy Guiliani), physically attacked the United States Capitol building in Washington, D.C., while Congress was certifying the election of the President-elect, Joseph Biden, Jr., in an effort to stop the certification process. The entire episode appeared to have been driven by of a kind of mass hysteria, reminiscent of the huge rallies that mobilized the German people on behalf of the National Socialists. The insurrection was in itself hideous enough, but particularly noteworthy in the context of this work was a underlying, ubiquitous, white supremacist theme: (1) Confederate battle flags were common, the owner of at least one of which was seen replacing the United States flag within the Capitol; (2) signs implying the six million Jews killed in the Holocaust were not enough; (3) a noticeable lack of protection for the Capitol, particularly when compared to the police and National Guard presence during anti-racism rallies held several months earlier; and (4) outspoken statements of white supremacy, for those who somehow missed the underlying white supremacist tenor of the crowd. A nearly ubiquitous absence of masks in the crowd (not to mention among Republican congresspersons within the Capitol) showed its utter disdain for the most basic preventative epidemiology. White supremacy, and the specious, irrational fears, anger, resentment, and hate that it represents, continued to be alive and well in the United States.

Ultimately, Trump's presidency was an extraordinarily odious time of overt racism and political conservatism not seen since the early—mid-20th century. The divisive fires of racism and intolerance were relentlessly stoked by a President whose indifference to the truth, as it is generally understood, was unprecedented. Instead of transparency and tolerance, he surrounded himself with sycophants and collaborators who embraced his white-supremacist world view, while he skillfully manipulated social media as a resonating chamber for his conspiracy-based ideas. In so doing, he exposed and reenergized deep veins of ignorance, racism, and intolerance, mostly dormant since the 1930s, that lie within with social fabric of United States. A full accounting of this President's activities and legacy, however, is not within the scope of this book.

Reflecting the polarity of American society, many initiatives have been undertaken to improve the lives of marginalized people, and in so doing, all of us. For example, the 1964 ambitious series of policy initiatives, legislation and programs spearheaded by President Lyndon B. Johnson, termed *The Great Society*, were designed to end poverty, reduce crime, abolish inequality and improve the environment. As a part of the Great Society, Johnson expanded the federal government's roles in education and health care as poverty reduction strategies. As a result of these policies, the percentage of Americans living below the poverty line declined from 22.2% to 12.6% (Califano, 2015). Additionally, Fernald et al. (2008) showed that by supplementing poor families' incomes there was an improvement in their children's cognitive and language skills within 18 months. The conclusion can be drawn that socio-cultural context has direct and almost immediate effects on cultural and academic performance.

Despite all its limitations and need for improvement, the United States is still, in the first part of the 21st century, a place where information is readily available, and people's voices can be quickly amplified via social media.

## Inequities in Modern Education

Education would seem to be the one obvious route to eliminating the vicious circles of poverty and discrimination we have documented here. Yet, as we have seen, systemic inequities are present in all aspects of education from preschool to college (see Chapter 5). Carnevale and Strohl (2013) reported that many college-qualified BIPOC students do not get college degrees due to limited higher education opportunities, reinforcing the legacy of white privilege. They reported that 82% of new white enrollments have gone to the 468 most selective colleges, but 72% of new Latinx enrollment and 68% of new Black enrollment have gone to two-year open-access schools. These separate higher educational tracks have real consequences for these students. Reardon et al. (2012) describe that between 1982 and 2004, the achievement gaps between black and white students narrowed in high schools but increased in college enrollments. They concluded that the growing racial enrollment gap is driven by changes in college admission, rather than the relative academic preparation of students.

Perhaps the most significant factor contributing to the racial stratification of higher education admissions is the weight of the standardized tests used for college admissions: SAT/ACT. Historically, these test scores were once thought of as aptitude tests, that is, proxies for the probability of success in college and after (itself, presumed to be a proxy for intellectual capacity), but in reality, these tests have always correlated more highly with demographics than with academic performance (Soares, 2007). In fact, over the last decade, race has become a higher predictor of SAT/ACT test scores than parent education or family income (Soares, 2020).

The SAT was created for the College Board as an intelligence test in 1926 by Carl Brigham of Princeton University. Originally there was a eugenic and ethnic diversity-limiting intent behind the SAT, which aimed to exclude Jews and BIPOC from Ivy League Universities (Soares, 2020). This agenda was realized by biased test-question selection that ultimately systemically discriminated against non-whites. Thus, experimental questions on which BIPOC (although the acronym had not yet been invented) outperformed whites were removed before administration of test questions because such shifts in scores would disrupt the expected normal distribution of SAT scores (Soares, 2020). In contrast to scores on standardized tests, BIPOC perform significantly better on curriculum-based achievement tests. Further undermining the significance of the SAT, regardless of the quality or type of school attended, high-school grade-point average (GPA) is the best predictor of freshman grades and long-term college outcomes such as cumulative GPA and four-year graduation (Atkinson & Geiser, 2009).

The self-serving myth perpetuated by the testing industry and those who benefit from the use of test scores in college admissions, has been that SAT/ACT scores predict college performance better than high-school grades. These disparities have been recognized, and recently there has been a move by some schools to eschew standardized tests as a requirement for admission. Still, SAT/ACT scores, which show Blacks and Latinx youths at the bottom, are still used by many universities and unquestionably perpetuate racial disparities.

Of particular relevance to this book are inequalities in science education. Data from the National Science Foundation (NSF, 2018) indicate that there are persistent trends of underrepresented BIPOC in the science and engineering workforce. In the latest NSF report, Latinx accounted for 6% of employment in science and engineering occupations, while they comprise 15 % of the US population age 21 and older. Similarly, African-Americans accounted for 5% of science and engineering employment, less than half of their share of the US population age 21 and older (12%). According to a recent Pew Research Center study (Funk & Parker, 2018), more than half of Science, Technology, Engineering, and Mathematics (STEM) workers believe that limited access to quality education is a major reason why Blacks and Latinx are underrepresented in STEM jobs. In contrast to underrepresentation in STEM fields, the National Association for the Advancement of Colored People (NAACP, 2019) reports that African-Americans and Hispanics make up approximately 32% of the US population, but they comprised 56% of all incarcerated people. These imbalances are reflective of systematic forces which limit access to power, higher economic, and social attainment. Such forces surely include access to quality education.

## Conclusion

*During our conversation about the hate murdering of Latinx, my son asked a very simple question: Why do people hate others based on their physical appearance? It would be nice to be able to lay this at the feet of thoughtless racism. But the truth is more incredible: it is* thoughtful *racism; a feeling so strong, so overpowering, so deeply held, and so resistant to information to the contrary, that the logic of scientific reasoning can actually be bent to serve it. People are fearful of that which is unfamiliar to them; which doesn't look like them. And yet . . .*

*And yet, our family is fortunate to have access to people who care enough about us to teach us and support our growth. Ron Killswarrior is more than just a friend to me. He is my oldest son's godfather, and a voice of wisdom in my life. During some very difficult moments, I have counted on Ron's insights and perspective. Ron also happens to be the first person who taught me about the injustices that were perpetrated to his people, the Lakota Sioux. When Ron took me to visit the site of Wounded Knee on the Pine Ridge Indian Reservation, and as we stood in the hundred-degree summer, I felt a cold chill down my spine as he described the murder of his ancestors. It was embarrassing to me that as a high school and college graduate in the United States, I was completely ignorant of the systemic atrocities that were committed to Native Americans. In the same way that a hammer can be a weapon and a building tool—biology education through the lenses of evolutionary biology is a tool for building a better way of thinking about our shared humanity.*

## Notes

1 A quote that belies the *post facto* claim that the American Civil War was actually about states' rights, and not about the institution of slavery.

2 The government's inability to keep its promises would have repercussions that are still felt today. For example, in places like the Mississippi Delta, 98% of black agricultural landowners have been dispossessed culminating in African-Americans losing over 12 million acres over the past century (see Agyeman & Boone, 2020; Newkirk, 2019).

3 These events inspired the 2022 movie "Till," a look at these events through the eyes of Till's mother, Mamie Till-Mobley. The film was selected as one of the best films of 2022 by the National Board of Review of Motion Pictures.

4 A powerful recounting of the gruesome 1947 lynching of a black man, Willie Earle, and the subsequent trial and exoneration of 31 guilty white men who carried out the deed is made by West (1955), in her essay "Opera in Greenville."

5 Trump characterized Mexicans hoping to immigrate into the United States as "drug dealers, criminals, and rapists," relentlessly playing to deep-seated white cultural fears of the "theft" of "their" America by BIPOC.

## References Cited

Agyeman, J., & Boone, K. (2020). *Land Loss Has Plagued Black America Since Emancipation—Is It Time to Look Again at "Black Commons" and Collective Ownership?* Retrieved August 21, 2020, from www.commondreams.org/views/2020/06/20/land-loss-has-plagued-black-america-emancipation-it-time-look-again-black-commons

American Experience. (2020). *The Trial of J.W. Milam and Roy Bryant*. Retrieved August 15, 2020, from www.pbs.org/wgbh/americanexperience/features/emmett-trial-jw-milam-and-roy-bryant/

Arce, J. (2019, August 6). *Donald Trump's Anti-Immigration Rhetoric Is Rooted in Racism*. Retrieved September 24, 2020, from https://time.com/5645501/trump-anti-immigration-rhetoric-racism/

Atkinson, R.C., & Geiser, S. (2009). Reflections on a Century of College Admissions Tests. *Educational Researcher, 38*(9), 665–676. https://doi.org/10.3102/0013189X09351981

Bancroft Davis, J.C. (1896/1997). *United States Reports*, Vol. 163, pp. 537–563. New York: Banks & Brothers, Law Publishers. [Reprinted in B. Thomas (Ed.), *Plessy v. Ferguson: A Brief History with Documents*, pp. 41–44, 48, 50. Boston: Bedford Books, 1997.]

Bloom, J., & Marin, W.E. (2016). *Black Against Empire: The History and Politics of the Black Panther Party*. Berkeley, CA: University of California Press.

Califano Jr., J.A. (2015). *The Triumph and Tragedy of Lyndon Johnson: The White-house Years*. New York: Simon and Schuster.

Carnevale, A., & Strohl, J. (2013). *Separate & Unequal: How Higher Education Reinforces the Intergenerational Reproduction of White Racial Privilege*. Georgetown Center on Education and the Workforce. https://cew.georgetown.edu/cew-reports/separate-unequal/#resources

Coaston, J. (2019, April 26). *Trump's New Defense of his Charlottesville Comments Is Incredibly False*. Retrieved September 24, 2020, from www.vox.com/2019/4/26/18517980/trump-unite-the-right-racism-defense-charlottesville

Cobb, J. (2018). *Voter-Suppression Tactics in the Age of Trump*. Retrieved September 5, 2020, from www.newyorker.com/magazine/2018/10/29/voter-suppression-tactics-in-the-age-of-trump

Collier, A.K. (2015, November 4). *The Black Panthers: Revolutionaries, Free Breakfast Pioneers*. Retrieved August 15, 2020, from www.nationalgeographic.com/culture/food/the-plate/2015/11/04/the-black-panthers-revolutionaries-free-breakfast-pioneers/

DEA. (2020). *Drug Enforcement Administration Staffing and Appropriations, FY 1972–2019*. Retrieved August 15, 2020, from www.dea.gov/staffing-and-budget

Durkee, A. (2020). House Subpoenas Postmaster General for USPS Documents Detailing Changes. *Forbes*. Retrieved August 31, 2020, from www.forbes.com/sites/alisondurkee/2020/08/31/house-to-subpoena-postmaster-general-louis-dejoy-usps-postal-service-documents-detailing-changes/?sh=262c83e2afef

FBI. (2016, May 18). *Emmett Till*. Retrieved August 15, 2020, from www.fbi.gov/history/famous-cases/emmett-till

Feldman, M., Pérez, M., Morris, K., Keith, D., Weiser, W.R., Waldman, M., Howard, E., & Morales-Doyle, S. (2020, May 27). *The Myth of Voter Fraud*. Brennan Center for Justice. Retrieved April 9, 2022, from www.brennancenter.org/issues/ensure-every-american-can-vote/vote-suppression/myth-voter-fraud

Fernald, L.C., Gertler, P.J., & Neufeld, L.M. (2008). Role of Cash in Conditional Cash Transfer Programmes for Child Health, Growth, and Development: An Analysis of Mexico's Oportunidades. *The Lancet*, *371*(9615), 828–837.

Forde, K.R., & Bowman, B. (2017, February 9). *Exploiting Black Labor after the Abolition of Slavery*. Retrieved August 15, 2020, from https://theconversation.com/exploiting-black-labor-after-the-abolition-of-slavery-72482

Funk, C., & Parker, K. (2018, January 9). *Racial Diversity and Discrimination in the U.S. STEM Workforce*. Retrieved August 22, 2019, from www.pewsocialtrends.org/2018/01/09/blacks-in-stem-jobs-are-especially-concerned-about-diversity-and-discrimination-in-the-workplace/

Glanton, D. (2020, June 14). *Column: Donald Trump Is Using America's Fear of Black Men to Destroy the Anti-Racism Coalition*. Retrieved September 24, 2020, from www.chicagotribune.com/columns/dahleen-glanton/ct-trump-black-men-fear-20200615-5yx3rzjuhfgilavqs65khpzy7q-story.html

Heynen, N. (2009). Bending the Bars of Empire from Every Ghetto for Survival: The Black Panther Party's Radical Antihunger Politics of Social Reproduction and Scale. *Annals of the Association of American Geographers*, *99*(2), 406–422.

Horwitz, T. (2014, December 1). *The Horrific Sand Creek Massacre Will Be Forgotten No More*. Retrieved August 10, 2020, from www.smithsonianmag.com/history/horrific-sand-creek-massacre-will-be-forgotten-no-more-180953403/

Huie, W.B. (1956). *Killers' Confession in Look (January 1956)*. Retrieved August 15, 2020, from https://famous-trials.com/emmetttill/1766-confession

Hutchinson, B., & Mallin, A. (2020, February 6). *Alleged El Paso Mass Shooting Suspect Hit with Federal Hate-Crime Charges*. https://abcnews.go.com/US/alleged-el-paso-mass-shooting-suspect-hit-federal/story?id=68807098

Immerwahr, D. (2019). *How to Hide an Empire: A History of the Greater United States*, 1st ed. New York: Farrar, Straus and Giroux.

Jacobson, M.F. (1998). *Whiteness of a Different Color: European Immigrants and the Alchemy of Race*. Cambridge, MA: Harvard University Press.

Kennedy, J.F. (1961). *Inaugural Address, January 20, 1961*. Retrieved October 9, 2020, from www.jfklibrary.org/learn/about-jfk/historic-speeches/inaugural-address

King, L. (2019, August 5). A Cultural Reckoning Over a President's Language as Critics Tie Shooting to Hateful Rhetoric. *Los Angeles Times*. www.latimes.com/politics/story/2019-08-04/el-paso-shooting-trump-rhetoric-white-supremacy

Kroll, A. (2020). *The Plot Against America: The GOP's Plan to Suppress the Vote and Sabotage the Election*. Retrieved October 9, 2020, from www.rollingstone.com/politics/politics-features/trump-campaign-2020-voter-suppression-consent-decree-1028988/

Lee, B. (2020, June 25). *Trump Once Again Calls Covid-19 Coronavirus The "Kung Flu."* Retrieved September 24, 2020, from www.forbes.com/sites/brucelee/2020/06/24/trump-once-again-calls-covid-19-coronavirus-the-kung-flu/

Lepore, J. (2018). *These Truths: A History of the United States*. New York: W.W. Norton & Company.

Levine, S. (2020). *Trump Admits He is Undermining USPS to Make It Harder to Vote by Mail*. Retrieved September 4, 2020, from www.theguardian.com/us-news/2020/aug/13/donald-trump-usps-post-office-election-funding

Liptak, A. (2019, June 27). *Supreme Court Leaves Census Question on Citizenship in Doubt*. Retrieved October 14, 2020, from www.nytimes.com/2019/06/27/us/politics/census-citizenship-question-supreme-court.html

Lyman, B. (2013). *Alabama Grants Posthumous Pardons to Scottsboro Boys*. Retrieved October 14, 2020, from www.usatoday.com/story/news/nation/2013/11/21/scottsboro-boys-pardoned/3662205/

Mancini, M.J. (1996). *One Dies, Get Another: Convict Leasing in the American South, 1866–1928*. Columbia, SC: South Carolina University Press.

McCaskill, N.D. (2020). *"It was Great": In Leaked Audio, Trump Hailed Low Black Turnout in 2016*. Retrieved September 5, 2020, from www.politico.com/news/2020/08/21/trump-black-voters-turnout-2016-398520

Minian, A.R. (2018). *Undocumented Lives: The Untold Story of Mexican Migration*. Cambridge, MA: Harvard University Press.

Moyer, J.W. (2015, August 21). Trump Says Fans Are "Very Passionate" After Hearing One of them Allegedly Assaulted Hispanic Man. *The Washington Post*. www.washingtonpost.com/news/morning-mix/wp/2015/08/21/trump-says-fans-are-very-passionate-after-hearing-one-of-them-alleged ly-assaulted-hispanic-man/

NAACP. (2019). *Criminal Justice Fact Sheet*. Retrieved August 22, 2019, from www.naacp.org/criminal-justice-fact-sheet/

Nellis, A. (2017, September 12). *Black Disparities in Youth Incarceration*. Retrieved August 16, 2020, from www.sentencingproject.org/publications/black-disparities-youth-incarceration/

Newkirk II, V.R. (2019). *The Great Land Robbery: The Shameful Story of How 1 Million Black Families Have Been Ripped from Their Farms*. Retrieved August 21,

2020, from www.theatlantic.com/magazine/archive/2019/09/this-land-was-our-land/594742/

NSF. Chapter 3 | Science and Engineering Labor Force. (2018, January). Retrieved August 22, 2019, from www.nsf.gov/statistics/2018/nsb20181/report/sections/science-and-engineering-labor-force/immigration-and-the-s-e-workforce

Ortega, B., & Bronstein, S. (2020). *Trump's False Claims on Mail-in Voting Do More to Harm Elections than Threat of Fraud, Experts Say*. Retrieved September 5, 2020, from www.cnn.com/2020/09/03/politics/election-threat-trump-mail-in-voting-claims-invs/index.html

Petrella, C., & Hasan Loggins, A. (2017, January 8). *"This Is a Country for White Men": White Supremacy and U.S. Politics*. Retrieved September 7, 2020, from www.aaihs.org/this-is-a-country-for-white-men-white-supremacy-and-u-s-politics/

Pettigrew, S. (2017). The Racial Gap in Wait Times: Why Minority Precincts Are Underserved by Local Election Officials. *Political Science Quarterly, 132*(3), 527–547.

Pilkington, E. (2021, December 24). Report Shows the Extent of Republican Efforts to Sabotage Democracy. *The Guardian*. Retrieved April 9, 2022, from www.theguardian.com/us-news/2021/dec/23/voter-suppression-election-interference-republicans

Prokop, A. (2019, July 11). *Trump's Census Citizenship Question Fiasco, Explained*. Retrieved October 14, 2020, from www.vox.com/2019/7/11/20689015/census-citizenship-question-trump-executive-order

Reardon, S.F., Baker, R., & Klasik, D. (2012). *Race, Income, and Enrollment Patterns in Highly Selective Colleges, 1982–2004*. https://cepa.stanford.edu/content/race-income-and-enrollment-patterns-highly-selective-colleges-1982-2004

Reidy, J.P. (2019). *Illusions of Emancipation: The Pursuit of Freedom and Equality in the Twilight of Slavery*. Chapel Hill: The University of North Carolina Press.

Risen, J., (2023). *The Last Honest Man: the CIA, the FBI, and the Kennedys – and One Senator's Fight to Save Democracy*. Boston: Little, Brown and Company.

Robinson, I. (2020). *Tennessee Passes Law That Would Punish Protestors by Taking Away Their Right to Vote*. www.theroot.com/tennessee-passes-law-that-would-punish-protestors-by-ta-1844813612?utm_medium=sharefromsite&utm_source=_facebook&fbclid=IwAR3QGtWs1fGDWuHQ8d96TNYovKRDiGBTWIeJL5pj217T7OCL4qzdk0ca2tc

Samalis-Aldrich, K., & Von Spakovsky, H.A. (2020). *Database Swells to 1,285 Proven Cases of Voter Fraud in America*. Retrieved October 1, 2020, from www.heritage.org/election-integrity/commentary/database-swells-1285-proven-cases-voter-fraud-america

Scottsboro: An American Tragedy. (2000). *American Experience | Scottsboro: An American Tragedy | Transcript*. Retrieved August 15, 2020, from www.shoppbs.pbs.org/wgbh/amex/scottsboro/filmmore/pt.html

Silverman, J. (2018). *The Most Dangerous Negro*. Brandeis University Writing Program. Retrieved February 15, 2023, from www.brandeis.edu/writing-program/write-now/2018-2019/silverman-jacob/index.html

Soares, J.A. (2007). *The Power of Privilege: Yale and America's Elite Colleges*. Stanford, CA: Stanford University Press.

Soares, J.A. (2020). *The Scandal of Standardized Tests: Why We Need to Drop the SAT and ACT*. New York: Teachers College Press.

Sterling, E. (2014). *Drug Laws and Snitching—A Primer | Snitch | FRONTLINE*. Retrieved August 16, 2020, from www.pbs.org/wgbh/pages/frontline/shows/snitch/primer/

Trefousse, H.L. (1997). *Andrew Johnson: A Biography*, Reissue ed. New York: W. W. Norton & Company.

Trump, D.J. (2019, June 27). *Seems Totally Ridiculous that Our Government, and Indeed Country, Cannot Ask a Basic Question of Citizenship in a Very Expensive, Detailed and Important Census, in this Case for 2020* [Twitter Moment]. https://twitter.com/realDonaldTrump/status/1144298731887628288

U.S. Department of Education. (2016, July 7). *Report: Increases in Spending on Corrections Far Outpace Education*. Retrieved August 16, 2020, from www.ed.gov/news/press-releases/report-increases-spending-corrections-far-outpace-education

Utley, R. (1963). *The Last Days of the Sioux Nation*. New Haven, CT: Yale University Press. https://time.com/5858169/americas-worst-moments/?utm_source=newsletter&utm_medium=email&utm_campaign=history&utm_content=20200703&xid=newsletter-history&et_rid=24816138

Vann Woodward, C. (1964, April). Plessy v. Ferguson: The Birth of Jim Crow. *American Heritage*, *15*(3).

Villa, L. (2020). *Here's Why Even the Suggestion of Sending Law Enforcement to the Polls Is Troubling*. Retrieved November 24, 2020, from https://time.com/5883990/voter-intimidation-trump-sheriff/

West, R. (1955). Opera in Greenville. In *A Train of Powder*, pp. 75–114. New York: Viking Press.

Wolf, R. (2020, October 13). *Supreme Court Temporarily Allows Trump Administration to End Census Head Count*. Retrieved October 14, 2020, from www.usatoday.com/story/news/politics/2020/10/13/2020-census-supreme-court-lets-administration-end-head-count-now/5975298002/

Zucchino, D. (2020). *Wilmington's Lie: The Murderous Coup of 1898 and the Rise of White Supremacy*. New York: Grove, Atlantic, Inc.

# 10

# Science and American Racism

*When I first saw his face, his appearance worried me. He had the epicanthic fold shape typical of many Asians. I quickly scanned my head to find where this "Chinito" had gotten his looks. "Who in the extended family has those features?" The nurse broke my concentration when she said something like: "He looks a little 'Downsie,' I am calling the doctor." At that moment it hit me like the weight of the world—my first-born child had an extra chromosome in his cells, trisomy 21. My wife let a single tear roll down her cheeks and told the nurse, "Say whatever you have to say, it will not make a difference! He is my son and I love him the same." I was still in shock and I had to find a way out of the room and told my wife that I had to move the car (the only reasonable excuse I could come up with at the time). I ran out of the hospital and got into my car. I pulled into an almost empty parking lot and began to cry desperately. This could not be happening! He was supposed to be everything I am, but better—the upgrade. In this boy all my hopes and dreams lived—he was going to be the stronger wrestler; the better musician; the greater scientist.*

*In the middle of this confusion, I kept thinking that this was not supposed to be my personal story. I sat desperate in my car and I began to shout at God or the Universe or whatever the hell made me feel like I was being punished. Enraged, I remember asking if I had not gone through enough shit in my life that now I had to deal with the immense challenges his condition would bring. How could this happen to me? This anger and sadness ruptured the capillaries on my face and I wanted to wake up from this moment of pain by punching my way out of the windows of my car. I remember feeling completely lost and then out of nowhere and for no logical reason, I heard the words used by my aunt in prayer, "give me the strength to say yes, whenever you call me, to do whatever you call me to do." Those words touched me gently like a passing*

DOI: 10.4324/9781003409144-14

*spring shower and in a single moment of transformation, I realized that I had been preparing for this. This was my calling, to raise this child. I understood that I was not being punished and instead I was about to embark on a new and wonderous journey that included a human with differences I did not predict.*

*Once I had composed myself, I closed my car door and marched upstairs to see our son—a perfect little boy, with Down syndrome. I asked my wife to let me hold him and without thinking about it, I looked into his eyes and made him the promises of always keeping him safe and to dedicate my life to raising him with all the opportunities possible. On that moment he became my young prophet, my "little Buddha." Eighteen years later, those who know Isaac, know that he is a bright and loving young man. He loves music and basketball. But more importantly, he has shaped the way I see the world, and the way his younger brother interacts with others—regardless of physical/mental differences. My son Isaac is a loved member of our family, and I would not want to change that. But there are others who, claiming to want the betterment of humanity, would have excluded him from life.*

*Historically, people have attempted to dictate who has the right to procreate and to be born. Someone like my son would have been excluded from society because of his genetic differences—robbing the world of everything he brings and the lessons he teaches. As we will see in this chapter, science has played an active role in empowering those looking for reasons to limit those deemed less worthy. Ultimately,* science *is always invoked.*

## Science and its Historical Contribution to Racial Oppression

With Chapters 8 and 9 having introduced American racial oppression, we now proceed to the role that science played and continues to play in that oppression. Scientists are fond of affirming that science transcends mere social issues like race; science, so it is claimed, is true no matter what color you are; no matter what faith you follow. And that might be true, except for this small detail: if science is to be explanatory, it requires humanist interpretation. Science does not have the absolute truth—demonstrable by universal proofs—that, for example, one finds in mathematics. Science should not be mistaken for math. It is ultimately a human endeavor, and is thus susceptible to human triumph and failure. Sagan (1997) highlights the close relationship between science and the social and political development of the nascent United States:

> *The American revolutionaries were creatures of the European Enlightenment which provides an essential background for understanding the origins and purpose of the United States. "Science and its philosophical corollaries," wrote the American historian Clinton Rossiter "were perhaps the most important*

*intellectual force shaping the destiny of eighteenth-century America. . . . Free inquiry, free exchange of information, optimism, self-criticism, pragmatism, objectivity—all these ingredients of the coming republic were already active in the republic of science that flourished in the eighteenth century.*

(p. 410)

While science has authority in our society, and those who know and use science have a particular and unique set of powerful tools at their disposal, science has never been carried out in a political vacuum, nor has it ever been freed from political and personal prejudices; indeed, since the very inception of racism described in the previous chapters, practitioners have sought to root their political and social agendas in a robust scientific foundation.

By the time of the legal emancipation of slavery (1865), racism was well-established in the United States, and race was accepted as scientifically well-supported. At that time and subsequently, the idea of biological races presupposed that people from different groups differ in meaningful ways, most notably in intelligence (Saini, 2019). How remarkable, that a complex, largely undefinable quantity like intelligence was reflected by superficial features like color! It was *convenient* that such differences could be ascertained by outward appearance and that these natural markers dictated where, in a presumed hierarchy of humanity, which rungs different races occupied (Wilkerson, 2020). Race-based discrimination was further supported by an interpretation of the Christian theology that described some people as not coming from God, but rather as children of Satan,[1] as was the case for Jews and other non-Christians (Fredrickson, 2002). However, since conversion to Christianity[2] was an option, a new way of justifying the oppression was needed and science obligingly filled the vacuum. As we will explore in this chapter, since the invention of "race," and ramping up in the latter half of the 19th century, "science" was always the ultimate and incontrovertible authority in support of racist beliefs.

A key component of racism has been, since its inception- and long before the science existed that could be marshalled to support it—the idea that Black people (and by extension all non-whites) were less civilized than whites, leading to the conclusion that they were more bestial than whites. A suite of racist tropes accompanied this original racist sin, including the idea that Blacks, being closer to nature, feel things more "emotionally" than the more cerebral whites and that, lacking civilization, are more attuned to the natural environment than whites. Later, science came to provide the intellectual authority for the legitimization of such racist ideas. Eventually the theory of evolution would play a role in supporting the idea that human races were biologically different and that human diversity followed a hierarchy with whites in the top position.

## Science and Racism—Science Gone Wrong

The birth of modern science, not by chance, coincided with the birth of racism, and whites embraced scientific approaches to explain the physical differences between peoples. During the Enlightenment (~17th and 18th centuries), the empirical evidence used for dividing human groups proposed that the outward physical characteristics of an individual reflected other unseen traits, like aptitude, potential, place in the social hierarchy, and even worth (see Saini, 2019).

### "Scientific" Racism

Swedish natural scientist Carolus Linnæus (1707–1778) is remembered for his hierarchical, binomial classification system of living organisms using empirical (observable) data; a system that became the basis for modern classification of living organisms (Fara, 2003) and is still in use (at least, informally). Linnæus' deep connections to Christianity and affinity to the oligarchic rule of Europe at the time, also meant that his system was at least partly based on the ideas in the Bible where humans were the pinnacle of the organismal hierarchy (the *Scala Naturae*, ladder of nature)[3] and all other (lower) species served the (uppermost) human cause (Howard, 1982). In the first edition of his *Systema Naturae* (Linné et al., 1964) he observed four varieties of humans based on biogeography and color, thus recognizing four varieties of *Homo sapiens: europeanus, asiaticus, americanus*, and *africanus*, which were said to be white, yellow, red, and black, respectively. With time, however, Linnaeus' system came to be larded with personal prejudices, so that by the definitive 1758 tenth edition (Gordh & Beardsley, 1999), without presenting any additional scientific evidence, Linnæus attributed personality characteristics to each: Europeans were depicted as white, gentle, and inventive; Asians, as sallow, avaricious, and easily distracted; Native Americans were reddish, stubborn, and easily angered; while Africans, as black, relaxed, childlike, and negligent (Charmantier, 2016).

Once these varieties, or "races" were defined, they became the basis of hierarchies which supported the politics of the time (Hällgren & Weiner, 2006). Linnæus' classification, developed in the mid-1700s with a non-evolutionary mindset and utterly unsupported racist stereotypes, became from the outset a powerful underpinning for racist generalizations. He pioneered defining the concept of race as applied to humans and his classification was used by many Europeans to validate the conquering and subjugation of members of the "lower" races, (that is, lower on the *Scala Naturæ*, as evidenced by their inferior civilizations) including, as we noted previously, the enforcement of slavery. Unambiguous echoes of the original racist overtones of the Linnæan classification, which provided the intellectual framework for the subjugation of non-white races for two centuries, are still audible today.

Along with the widespread use of Linnaeus' classification, western scientists attempted to investigate the race hierarchy scientifically. American craniologist Samuel Morton (1799–1851) collected and measured hundreds of human skulls to infer the brain size of the racial groups; inferring that brain size was a proxy for intellect. Morton's conclusions, happily concordant with his expectations, were that whites had the largest skull sizes and therefore, the highest intelligence; while Africans had the smallest skull sizes and, thus, the lowest intelligence. These results were naturally cited to maintain racial segregation and Black slavery in the United States (Mitchell, 2018). Gould (1996) demonstrated that Morton's methods were not statistically meaningful, and Mitchell (2018) more recently argued that Morton's work was irreparably flawed because it lacked body-size data, since in humans, as in all vertebrates, bigger body sizes have bigger skulls—regardless of race. In the end, however, it has been known for many years that brain (or its proxy, skull size)-to-body size ratios correlate with intelligence at only the coarsest level, far too blunt an instrument to distinguish "races" within a single species.

Slave owner, American surgeon, and anthropologist Josiah Clark Nott (1804–1873) used his scientific reputation to defend slavery (Smith, 2014). Nott believed in the multiple geographic and phylogenetic origins of races, a now largely discredited viewpoint, although in Knott's time, very little was understood about human origins. Addressing the Southern Rights Association of Mobile in 1850, he channeled the *Scala Naturæ* by identifying Blacks as "the lowest point in the scale of human beings (Nott, 1851, p. 330)." Secure in that world view, he concluded that if Black and White races mixed, it would lead to extinction through degeneration.

Swiss-born American biologist and geologist Louis Agassiz (1807–1873), among the most famous scientists of his generation, lent his considerable prestige to the notion that separate races originated in different places and possessed distinct characters and intellectual abilities (Saini, 2019).[4] Along with Agassiz and Nott, British Egyptologist George Gliddon (1809–1857) published the book *Types of Mankind* (1854) with Agassiz and Nott. It sold three and a half thousand copies in the first four months and ran to ten printings over 17 years. The book claimed to provide scientific "proof" of the inferiority of Black people, an argument conveniently timed in its appeal to Southerners (Smith, 2014). Gliddon and colleagues' argument was that God had created Blacks separately and inferiorly. The book provided pictorial testimony in which black skulls where likened to those of chimpanzees and gorillas, as well as supposed ancient writings that showed Blacks as slaves and servants since antiquity.

Although Charles Darwin's (1809–1882) ideas ultimately would be contorted to explain racial differences and the oppression of non-whites, in his book *The Descent of Man*, (1889), Darwin swam against the tide of popular and scientific thought by presciently espousing racial unity by proposing that all humans share a most recent, *human*, common ancestor,

> *Now when naturalists observe a close agreement in numerous small details of habits, tastes, and dispositions between two or more domestic races, or between nearly-allied natural forms, they use this fact as an argument that they are descended from a common progenitor who was thus endowed; and consequently that all should be classed under the same species. The same argument may be applied with much force to the races of man. As it is improbable that the numerous and unimportant points of resemblance between the several races of man in bodily structure and mental faculties (I do not here refer to similar customs) should all have been independently acquired, they must have been inherited from progenitors who had these same characters.*
>
> (p. 179)

Note that Darwin's argument, for a single ancestor of humanity that must itself have been human, is a fundamentally evolutionary one, underlining the key connection between races and evolutionary biology. Yet, neither Darwin's stature nor the force of his argument impeded the acceleration of the scientific-race behemoth. Instead, ironically, Darwin's theory of evolution by natural selection (1859) instead became the justification for some of the most virulent and pernicious racism in history (Fuentes, 2021).

## Eugenics

If science could describe the natural world and validate social hierarchies, could science not be used to make a better world? In the late 19th century, British biologist Sir Francis Galton (1822–1911), proposed to apply his cousin Charles Darwin's ideas of evolution by natural selection to humans, much as had been carried out with domesticated animals for centuries by breeders. He termed this proposal "eugenics."[5] Galton (1883) defined eugenics as:

> *The science of improving stock, which is by no means confined to questions of judicious mating, but which, especially in the case of man, takes cognizance of all influences that tend in however remote a degree to give to the more suitable races or strains of blood a better chance of prevailing speedily over the less suitable than they otherwise would have had.*
>
> (p. 17)

Studying the ancestry of 605 nobleman who lived between 1453 and 1853, he determined that one in six was related to another member of the same group. Ignoring the environmental factors (wealth, education, and position in the rigidly socially stratified Victorian and Edwardian England in which he lived) he concluded that the ingredients for nobility, reflecting genetic superiority, must be the result of superior breeding (Saini, 2019). He wondered if it was possible for all humanity to gain the well-bred qualities he so admired in nobility, and this in turn let him to attempt to perfect the human race, by eliminating deficient examples of their "genes" from the human genetic pool by artificial selection. We note that his ideas were very much in accord with the mutationist school of evolution, in which an organism's fate was presumed to be almost solely controlled by its genetic composition, and in which genetic composition was presumed to be the deciding factor in natural selection.

Modern, scientific, "Race" self-improvement through artificial selection; the idea was catchy, and developed a significant following. American naturalist David Starr Jordan (1851–1931), first, a young President of Indiana University; later the founding President of Stanford University, was an early proponent of eugenics and, with it, naturally, the "racial" superiority of whites. He stated, "To say that one race is superior to another is merely to confirm the common observation of every intelligent citizen" (as quoted in Burns, 1953, p. 63). Jordan became an articulate and relentless proponent of eugenics (Miller, 2020). He went on to become a founding member of the Human Betterment Foundation, devoted to the promotion of legislation that would allow for compulsory forced sterilization (Johnsson, 2016).

Additional racist views were declared and supported by prominent American eugenicists such as the psychologists Henry Herbert Goddard (1866–1957), who concluded that the very high percentages—in the 80s—of the immigrant groups that he tested, mainly from eastern Europe, were "feeble-minded;" Lewis Terman (1877–1956), an early developer and proponent of IQ testing; and Education and Psychology Professor, later Chancellor of USC, Rufus Bernhard von KleinSmid (1875–1964), another founding member of the Human Betterment Foundation, who was famously quoted as stating "the application of the principles of Eugenics to organized society is one of the most important duties of the social scientist of the present generation." These people were among many who deemed that "inferior races," were a drain on the economic, political and moral health of American life (Markel, 2018). American eugenicists imagined a "blond and blue-eyed" Nordic ideal and anyone who did not resemble this model including all non-whites and immigrants (some of whom would be considered "white" at other times in other places), were targets for elimination from the population (Black, 2012). Eugenics offered Americans scientific justification for cleansing the body politic of undesirables.

Unsurprisingly, the eugenics movement found its way to Germany. The famous German zoologist and embryologist Ernst Haeckel (1834–1919) recognized 9 hierarchically arranged *species* of human beings naturally; the German or Nordic species was at the top (Richards, 2008). He too became an ardent supporter of eugenics by forthrightly noting that "politics is applied biology" (Black, 2012). Shortly after his death, his views came to be admired, studied, and ultimately emulated by the National Socialist German Workers Party (Nazi) (Gasman, 1971) and, in the name of retaining the purity of the "Aryan race" led directly to the mass sterilizations and murders of the Holocaust. The state-supported adoption of eugenics by the Nazis was enviously viewed by the superintendent of the Western State Hospital in Richmond, Virginia, Joseph S. DeJarnette (1866–1957), who stated that the German sterilization law that had come into effect on 1 January of 1934 was a demonstration that, "The Germans are beating us at our own game" (as quoted in Kevles, 1995, p. 116).

Eugenics had a genetics-based view, grounded in rigid biological determinism. The idea was that people had to become who their genes said they were; that environmental influences were ultimately trumped by genetic make-up. This perception of biological determinism, where genes were the totality of the individual, was largely led by American educator, eugenicist, and sociologist Harry Hamilton Laughlin (1880–1943). As the Superintendent of the Eugenics Record Office at the Cold Spring Harbor Laboratory in New York, Laughlin directed field workers to register and document those who were deemed feebleminded. Feeblemindedness, however, was a truly subjective quantity, and people were listed as such based on a wide variety of fluctuating reasons (Black, 2012).

The goal of purifying America's population developed into an alliance between biological racism and elitist power. Because the eugenic visions were well-funded by wealthy donors such as Andrew Carnegie and John D. Rockefeller, their reach was massive and politically impactful (Black, 2012). Laughlin provided extensive statistical testimony to the United States Congress in support of the Johnson-Reed Immigration Act of 1924, when he revisited H.H. Goddard's studies which had suggested that there was excessive insanity among immigrants from southern and eastern Europe. For the purpose of limiting people already within the United States, Laughlin wrote the *Model Eugenical Sterilization Law*, a model act for compulsory sterilization which would be applied in many states. Subjects for eugenic sterilization included "feeble-minded," the equally nebulous "insane," criminals, epileptics, alcoholics, blind persons, deaf persons, deformed persons, and indigent persons. Not unexpectedly, non-whites, along with white immigrants from the wrong places (eastern Europe), were disproportionately identified within these categories, serving racist intents. For instance, Price et al. (2020) report that after analyzing the evidence, North Carolina's eugenic sterilization program was racially biased and targeted black Americans with the intent to "breed (them) out."

In the United States the eugenics program included several large-scale interventions which combined the encouragement for higher rates of reproduction for racial groups deemed eugenically superior (White Anglo-Saxon Protestants), with strict racial and immigration quotas, as well as the forced sterilization of citizens with undesirable traits (Markel, 2018). Twenty-seven states enacted mandatory sterilization laws, along with marriage prohibition laws to stop race mixing. Even the US Supreme Court legitimized eugenics in the Buck v. Bell decision of 1927 (Black, 2012). Conservatively, an estimated 60,000 Americans were coercively sterilized. Although eugenics wore the mantle of respectable science, its true nature was simply racism, ethnic fear, and academic elitism (Black, 2012).

## Other Ways in Which Science was Racist

Aside from the very visible face of the eugenics program in the United States, there have been many other instances where science has been utilized to validate racist behavior. For example, James Marion Sims (1813–1883) a famous American surgeon of the 19th century, perfected the method for repairing vesicovaginal fistulas (Lynch, 2020). Shockingly, it has been reported that Sims conducted experimental surgery on at least 10 enslaved women without anesthesia, an act he did not perform to any white women (Gamble, 1997). Since these women were property, there was no consent for the surgeries, and Sims's decision to not use anesthesia was based on the belief that Black people were intrinsically different and did not experience pain like white people did (Axelsen, 1985). Additionally, Sims was convinced that the skulls of Black people grew too quickly around their brain, making them less intelligent than white people (Holland, 2017).

Perhaps the most infamous and best-documented racist science was conducted by US Surgeon General Thomas Parran (1892–1968) who oversaw the notorious 1932 Tuskegee biomedical experiment in which researchers purposefully refused black patients the antibiotics that would cure them, to track the effects of syphilis and document the degradation of the patients' internal organs and eventual deaths. In 1946 he also approved unethical experiments in Guatemala, where female sex workers, prison inmates, mental patients and soldiers were unknowingly infected with syphilis or gonorrhea (Reverby, 2011).

Cornelius P. Rhoads Packard (1898–1959) was another high-profile American physician involved in racist and unethical human experimentation. In a private letter from November 1931, Packard expressed the hatred he had for the people who inhabited the island where he was working,

> Porto [*sic*] Ricans, are beyond doubt the dirtiest, laziest, most degenerate and thievish race of men ever inhabiting this sphere. What the island needs is not public health work but a tidal wave or something to totally exterminate the population. I have done my best to further the process of extermination by killing off 8 (by injection with cancer cells; as quoted in Lederer, 2002, p. 720).

Rhoads would later say that the letter was a joke. Moreover, after an investigation from the Rockefeller Institute which sponsored him, he was found not guilty of any medical wrongdoing. Regardless, at a minimum, the letter suggests a deep well of racism.

Recently attention has been given to Henrietta Lacks (1920–1951). Lacks was a black woman whose cells were used to form the HeLa cell line, which has been used extensively in medical research (Skloot, 2010). Months before her untimely death from cervical cancer at 31 years old, doctors at the Johns Hopkins Hospital in Baltimore, one of only a few places that provided medical care to Black people at the time, took samples of her cancerous cells while diagnosing and treating her disease. Without Lacks' knowledge or consent, sample tissues were given to a researcher, Dr. George Otto Gey (1899–1970). In the laboratory, her cells had an extraordinary capacity to survive and reproduce and came to be known as "immortal." Gey and his team shared the cells widely with other scientists, and these cells became a staple of cellular biological research. In many ways, the use of HeLa cells supports much of modern medicine since these have been involved in vital breakthroughs in many fields, including *in vitro* fertilization, cancer, immunology and infectious disease. In fact, a recent application of HeLa cells was in the research for vaccines against COVID-19 (Nature, 2020).

The story of Henrietta Lacks illustrates the entrenched racial inequities in the US research and health-care systems. Although billions of dollars have been generated thanks to her cells, none of the biotechnology or other companies that profited from using the HeLa cell line provided any compensation to her family. Physicians and scientists continually failed to ask her family for consent as they divulged her medical records, and even published her cells' genome online (Nature, 2020). We have no examples of similar stories involving white people. It is likely not coincidence that, like Lacks, DNA samples that had been donated by the Havasupai Tribe for genetic research on type-2 diabetes in 1989 were in fact being used without permission in genetic studies unrelated to diabetes, by researchers at Arizona State University. In 2004, the Tribe filed and won a lawsuit against the Arizona Board of Regents and Arizona State University (Garrison, 2013).

### Racial Stereotypes

Racial stereotypes, many of them grotesque, characterize racist thinking. Despite their absurdity, as well as a glaring absence of evidence for their support, they persist. For example, the threadbare racist canard depicting black people as more sexually aggressive than whites was revived as recently as the early 21st century. In a now retracted paper, Rushton and Templer (2012) suggested that darker pigmented humans have higher amounts of aggression and sexual activity than lighter pigmented individuals. Similar beliefs about the physiological differences between groups of humans is also held by medical professionals. Hoffman et al. (2016) found that the majority of the white medical students and residents that participated in their study reported believing that there are real racial differences in skin thickness and pain reception between blacks and whites. These beliefs in turn, played a deleterious role in treatment recommendations. For example, Black medical patients receive more amputations that white patients (Stapleton et al., 2018), a statistic that might derive its horror from more than just a misunderstanding of pain reception.

## Modern Ideas of Intelligence and Race

Racist ideology and behavior encompassed many scientific fields including psychology. Arthur Jensen (1923–2012) a professor of educational psychology at the University of California, Berkeley, was a major proponent of biological determinism and proposed that genetics were fundamental in behavioral traits, such as intelligence and personality. In a published article (Jensen, 1969) he concluded, among other things, that Head Start programs designed to boost Black IQ scores were a failure, and that this was likely never to be remedied, because, according to him, 80% of the variance in IQ in the population studied was the result of genetic factors. For his work, Jensen received $1.1 million from the Pioneer Fund, an organization frequently described as racist and white supremacist in nature (Saini, 2019). Gould (1996) offers lucid criticism of Jensen's work specifically by pointing out two major flawed points, the reification of intelligence as a single, measurable entity; and the misunderstanding that evolution happens as unilinear progress, on an observable scale, ascending from primitive to advanced representing the best way of ordering variation (that is, the venerable, discredited *Scala Naturae*).

The racist baton carried by Jensen was picked up by Richard J. Herrnstein and Charles Murray in their highly publicized (and criticized) book, *The Bell Curve* (1994). In the book, the authors report on racial differences in IQ score, with Black people in the United States scoring lower on average than white people. For these authors, race is a meaningful indicator of intelligence. The book proposes that the United States should stop trying to improve the material living standards of poor people because doing so encourages poor, low-IQ women to have more children. Gould (1996) provides again scathing criticism for *The Bell Curve*, by pointing out four unsupported and false assumptions made by the authors: (a) intelligence is reducible to a single number; (b) it is capable of rank ordering; (c) it is primarily genetically based, and (d) as such it is essentially immutable.

Alvin Poussaint (1934–), a professor of psychiatry at Harvard, further critiqued Herrnstein and Murray's conclusions by noting that the racially based groups they identified are culturally and socially biased. Poussaint argued that hundreds of years of legal slavery, when Blacks were not allowed to learn to read and write, followed by a long history of racial segregation, when people were separate and not equal, have rendered whites and Blacks not comparable by the same indices. For Poussaint, the only way to show that genes determined intelligence, was to find a group of whites who went through 250 years of slavery and 100 years of segregation (Spencer, 1995).

Because racism seems to tap deeper feelings than the intellect, when racist science is logically dismantled, it finds ways of surviving. Murray mentored Jason Richwine (1982–) whose Harvard doctoral dissertation (Apocrypha, 2009) argued that Latinx immigrants have lower IQs than what he called "native whites," and that, in an echo of the impurity arguments of the late 1800s, their migration to the United States is lowering the collective IQ of Americans. Richwine concluded, "No one knows whether Hispanics will ever reach IQ parity with whites, but the prediction that new Hispanic immigrants will have low-IQ children and grandchildren is difficult to argue against" (p. 66). Ultimately built upon racist, discredited tropes (a correlation between "race" and intelligence), Richwine's dissertation lacked analytical rigor, included specious use of data, and used gross generalization rather than disciplined scholarship. For example, in selecting the group "Hispanics," Richwine does not account for the enormous variation within the population which includes descendants of indigenous peoples, enslaved Africans, and European immigrants.

Regardless of the academic flaws of this work, Richwine was hired by the conservative thinktank the Heritage Foundation where he produced a controversial report that criticized the potential cost of immigration reform. Recently, he has been associated with the Center for Immigration Studies, a group that informed President Trump's views on immigration (Reston, 2017).

Finally, New York University professor of Politics and Public Policy, Lawrence Mead (2020), writing for the journal Society, claimed that because most poor people in America are Blacks and Latinx, and thus it can be said that because they did not come from Europe, American poverty is due to non-European origins. The article was subsequently retracted, but it clearly tapped a deep vein of antagonism, rather than the Enlightenment rationality upon which America was founded.

## Current Events and Trends

The impact of the COVID-19 pandemic was felt across society, but the most significant negative effects were (and are) felt by BIPOC. Racial disparities show up in a variety of measurable outcomes including rate of infection, mortality rate and negative economic impact—all which disproportionally impact racially minoritized people (Wood, 2020; Morning Edition, 2020). The data show that BIPOC bear the heaviest burden of COVID-19 at every stage, from risk of exposure, to access to testing, to severity of the illness and ultimately, to death. These differences are not due to biological or racial differences, but rather to systemic social disparities (Wood, 2020).

Other ways in which scientific racism shows up include the collection of DNA from immigrants at the Mexican border, but not of Europeans coming via airplanes. Under the specious pretext of targeting criminals, DNA is collected from all immigrants in detention at the Mexican border, including children. That there might be a DNA marker for criminality is an echo of the Bad Old Days of eugenics. Even if the rationale is not that now-discredited science, knowing the history of xenophobia and racism in the United States does not alleviate the suspicion that the DNA information can be seen as a precursor to deploying surveillance technology to vulnerable communities in the United States and ultimately be used as a tool for full-scale surveillance (Kennedy, 2020; Hackman, 2020). In fact, the data show that immigrants commit fewer crimes than native-born Americans (Light & Miller, 2018); yet with the history of misuse of biological data, it is not surprising that some people are worried about the abuse of this nascent DNA library.

Other echoes of the eugenics movement can be heard today. A whistleblower complaint filed on September 14, 2020, alleges that immigrants held in custody at a private detention facility in Georgia contracted by US Immigration and Customs Enforcement (ICE) feared for their lives due to negligent COVID-19 practices by facility staff. The allegations included reports of unwanted hysterectomies performed on detained women (Treisman, 2020).

Even when medical negligence or misconduct are not directly present, the current general treatment of immigrants, the highly charged public discourse, and oppressive policies have detrimental impact on the lives of immigrant communities. The current immigration prosecution generates fear in immigrant communities across the country, including US-born children and legal immigrants. A report from the Migration Policy Institute (Capps et al., 2020) concluded that the majority of students from immigrant families surveyed feared that someone close to them would be arrested and deported and more than half reported symptoms of mental-health conditions such as anxiety, post-traumatic stress disorder (PTSD), or depression at levels high enough to warrant treatment. Childhood trauma, or adverse childhood experiences (ACEs; experiences that would be common among refugee families) are associated with higher rates of heart disease, stroke, cancer, and diabetes in middle-aged patients (Felitti et al, 2019). Moreover, there is an undeniable relationship between childhood trauma and the onset of learning and behavioral issues (Burke et al., 2011). These are the kind of data neglected and ignored by the people who use biological determinism to argue for racial differences (see earlier section of this text).

## Conclusion

Science has not been an innocent bystander; its prestige and intellectual power have been used, again and again, to attempt to provide definitive support for racism. Science has been at best, insensitive to, and at worst, complicit in maintaining the white center of power. Science education must play a pivotal role in empowering a new generation to fight against racism and the ideologies which have for so long limited the success of some groups of Americans.

*My son Isaac has been a blessing for me and my family. In many ways, his unique genetic combination makes him truly special. At the beginning, the response from people was the usual "I'm sorry," to which I would normally respond, "I said he has Down syndrome, I didn't say he was dying. He will have challenges, but our family is ready to help each other through them."*

*With Isaac's birth, a new paradigm was required—after all, how could I love this child with cognitive disabilities who would struggle with many aspects of life, if the things I valued the most were intellectual capacity and academic advancements? I needed a new viewpoint; a perspective that would enlarge my understanding of humanity and the very meaning of life. It was his birth that began changing my heart about my own future and turned my ideas about the purpose of my life. After some struggle with these new perspectives, I decided to earn a high-school teaching certificate where I could use my passion and knowledge of science into a rewarding experience for others. Even after it was complete, I was hesitant to leave my job at an Ivy League university and take a significant pay cut in order to try something where I would be a rookie. Once I walked into room 132 where I once sat as a science student, however, hesitation vanished—I was home, and this time I was armed with knowledge about science, the world, and myself. For me and for the many classes of students that followed, science has been more than a ticket out of more than monetary poverty—it has provided the skills, critical thinking, and confidence to analyze and reject unjust social constructs.*

## Notes

1 *The English Standard Version Bible,* 2009. 1 John 3:10, "By this it is evident who are the children of God, and who are the children of the devil: whoever does not practice righteousness is not of God, or is the one who does not love his brother."

2 John 1:12–13, "But to all who did receive him, who believed in his name, he gave the right to become children of God, who were born, not of blood nor of the will of the flesh nor of the will of man, but of God."

3 The *Scala Naturae* in its original Christian incarnation was seen as a linear continuum from the "bestial" through the "spiritual." Animals were of course bestial and completely of this world; angels, and ultimately God, were spiritual, inhabiting an otherworldly spiritual perfection. Humans were thought to possess a mixture of these qualities, and were therefore torn, so to speak, between their bestial and spiritual impulses; they would thus be located along the continuum of the *Scala Naturae* somewhere between and the bestial and the spiritual. Nineteenth- and early 20th-Century evolutionists, observing morphological changes through time in lineages of organisms (seen in the fossil record), thought of evolution as linear, generally, increasing in complexity, and thus improving (which was thought to mean becoming "better adapted"), much as improvement was sought and hypothesized to occur in human societies, either through the introduction of humane laws or, in some quarters, through eugenics. Because the record of evolution was mistakenly perceived as linear, evolutionists comfortably—and wrongly—overlaid their observations from the fossil record onto the venerable *Scala Naturae,* interpreting it as the shape of organic evolution. This error in understanding has unfortunately become an icon for organic evolution, a point wittily made by S.J. Gould (1989).

4 This was an early expression of the multi-regional hypothesis of human origins, as opposed to the hypothesis of a single African origin for humanity. In this early incarnation, the multi-regional hypothesis had pernicious racial overtones, which remain stumbling blocks for modern scientists who still promulgate a less-overtly racist version of the multi-regional hypothesis.
5 The word derives from the Greek root, "*eu*- (meaning "true" or "good" and *gen*- (meaning "origin;" a reference to genes).

## References Cited

Apocrypha. (2009). *The English Standard Version Bible: Containing the Old and New Testaments with Apocrypha*. Oxford: Oxford University Press.

Axelsen, D.E. (1985). Women as Victims of Medical Experimentation: J. Marion Sims' Surgery on Slave Women, 1845–1850. *Sage: A Scholarly Journal on Black Women, Fall*, 2(2), 10–13.

Black, E. (2012). *War Against the Weak: Eugenics and America's Campaign to Create a Master Race,* 2nd ed. Washington, DC: Dialog Press.

Burke, N.J., Hellman, J.L., Scott, B.G., Weems, C.F., & Carrion, V.G. (2011). The Impact of Adverse Childhood Experiences on an Urban Pediatric Population. *Child Abuse and Neglect, 35*(6), 408–413. https://doi.org/10.1016/j.chiabu.2011.02.006

Burns, E.M. (1953). *David Starr Jordan: Prophet of Freedom*. Stanford: Stanford University Press.

Capps, R., Berger Cardoso, J., Brabeck, K., Fix, M., & Ruiz Soto, A.G. (2020). *Immigration Enforcement and the Mental Health of Latino High School Students*. www.migrationpolicy.org/research/immigration-enforcement-mental-health-latino-students

Charmantier, I. (2016). *Linnaeus and Race*. Retrieved January 14, 2021, from www.linnean.org/learning/who-was-linnaeus/linnaeus-and-race

Darwin, C. (1859). *On the Origin of Species by Means of Natural Selection, or the Preservation of Favoured Races in the Struggle for Life*. Retrieved June 22, 2020, from http://darwin-online.org.uk/contents.html#origin

Darwin, C. (1889). *The Descent of Man, and Selection in Relation to Sex*. New York: D. Appleton and Company. http://darwin-online.org.uk/converted/pdf/1889_Descent_F969.pdf

Fara, P. (2003). *Sex, Botany & Empire: The Story of Carl Linnaeus and Joseph Banks*. New York: Columbia University Press.

Felitti, V.J., Anda, R.F., Nordenberg, D., Williamson, D.F., Spitz, A.M., Edwards, V., Koss, M.P., & Marks, J.S. (2019). Relationship of Childhood Abuse and Household Dysfunction to Many of the Leading Causes of Death in Adults. The Adverse Childhood Experiences (ACE) Study. *American Journal of Preventive Medicine, 56*(6), 774–786. https://doi.org/10.1016/j.amepre.2019.04.001

Fredrickson, G. (2002). *Racism: A Short History*. Princeton; Oxford: Princeton University Press. https://doi.org/10.2307/j.ctvc779fw

Fuentes, A. (2021). "On the Races of Man": Race, Racism, Science, and Hope. In J. Desilva & J. Browne (Eds.), *A Most Interesting Problem: What Darwin's Descent of Man Got Right and Wrong About Human Evolution*, pp. 144–161. Princeton, NJ: Princeton University Press.

Galton, F. (1883). *Inquiries into the Human Faculty & Its Development*. JM Dent and Company.

Gamble, V.N. (1997, November). Under the Shadow of Tuskegee: African Americans and Health Care. *American Journal of Public Health, 87*(11), 1773–1778.

Garrison, N.A. (2013). Genomic Justice for Native Americans: Impact of the Havasupai Case on Genetic Research. *Science, Technology & Human Values, 38*(2), 201–223. https://doi.org/10.1177/0162243912470009

Gasman, D. (1971). *The Scientific Origins of National Socialism: Social Darwinism in Ernst Haeckel and the German Monist League*. New York: Science History Publications.

Gliddon, G., Nott, J.C., & Agassiz, L. (1854). *Types of Mankind: Or, Ethnological Researches, Based Upon the Ancient Monuments, Paintings, Sculptures, and Crania of Races, and Upon Their Natural, Geographical, Philological and Biblical History*. Philadelphia, London: Lippincott Gramoo & Co., Trubner & Co.

Gordh, G., & Beardsley, J.W. (1999). Taxonomy and Biological Control. In T.S. Bellows & T.W. Fisher (Eds.), *Handbook of Biological Control: Principles and Applications of Biological Control*, pp. 45–55. London: Academic Press. ISBN 978-0-12-257305-7.

Gould, S.J. (1989). *Wonderful Life*, p. 350. New York: W.W. Norton Co.

Gould, S.J. (1996). *The Mismeasure of Man*. New York: Norton & Company.

Hackman, M. (2020, March 6). *Trump Administration to Collect DNA From Immigrants Taken into Custody*. Retrieved September 24, 2020, from www.wsj.com/articles/trump-administration-to-collect-dna-from-immigrants-taken-into-custody-11583470503

Hällgren, C., & Weiner, G. (2006). Out of the Shadow of Linnaeus: Acknowledging their Existence and Seeking to Challenge, Racist Practices in Swedish Educational Settings. *Education-Line*. Retrieved September 13, 2020, from www.leeds.ac.uk/educol/documents/157423.htm

Herrnstein, R.J., & Murray, C. (1994). *The Bell Curve: Intelligence and Class Structure in American Life*. New York: Simon and Schuster.

Hoffman, K.M., Trawalter, S., Axt, J.R., & Norman Oliver, M. (2016). Racial Bias in Pain Assessment and Treatment Recommendations, and False Beliefs about Biological Differences between Blacks and Whites. *Proceedings of the National Academy of Sciences, 113*(16), 4296–4301.

Holland, B. (2017, August 29). *The "Father of Modern Gynecology" Performed Shocking Experiments on Slaves*. Retrieved September 17, 2020, from www.history.com/news/the-father-of-modern-gynecology-performed-shocking-experiments-on-slaves

Howard, J. (1982). *Darwin: A Very Short Introduction*. New York: Oxford University Press.

Jensen, A. (1969). How Much Can We Boost IQ and Scholastic Achievement? *Harvard Educational Review, 39*(1), 1–123.

Johnsson, L. (2016). *Guest Opinion: The Inconvenient Truth about David Starr Jordan*. Retrieved September 17, 2020, from https://paloaltoonline.com/news/2016/02/19/guest-opinion-the-inconvenient-truth-about-david-starr-jordan

Kennedy, M. (2020, March 6). *Trump Administration Poised to Start Collecting DNA From Immigration Detainees*. Retrieved September 24, 2020, from www.npr.org/2020/03/06/812940401/trump-administration-poised-to-start-collecting-dna-from-immigration-detainees

Kevles, D. (1995). *In the Name of Eugenics: Eugenics and the Use of Human Heredity*. Berkeley: University of California Press.

Lederer, S.E. (2002). "Porto Ricochet": Joking about Germs, Cancer, and Race Extermination in the 1930s. *American Literary History, 14*(4), 720–746.

Light, M.T., & Miller, T. (2018). Does Undocumented Immigration Increase Violent Crime? *Criminology, 56*(2), 370–401.

Linné, C. V., Engel-Leboer, M. S. J. & Engel, H., eds. (1964). *Carolus Linnaeus Systema naturae, Facsimile of the 1st ed*. Nieuwkoop: B. de Graff.

Lynch, S. (2020, June 20). Fact Check: Father of Modern Gynecology Performed Experiments on Enslaved Black Women. *USA Today*. www.usatoday.com/story/news/factcheck/2020/06/19/fact-check-j-marion-sims-did-medical-experiments-black-female-slaves/3202541001/

Markel, H. (2018). *Column: The False, Racist Theory of Eugenics Once Ruled Science. Let's Never Let That Happen Again*. Retrieved September 19, 2020, from www.pbs.org/newshour/nation/column-the-false-racist-theory-of-eugenics-once-ruled-science-lets-never-let-that-happen-again

Mead, L.M. (2020). Retracted Article: Poverty and Culture. *Society*. https://doi.org/10.1007/s12115-020-00496-1

Miller, L. (2020). *Why Fish Don't Exist: A Story of Loss, Love, and the Hidden Order of Life*. New York: Simon and Schuster.
Mitchell, P.W. (2018). The Fault in His Seeds: Lost Notes to the Case of Bias in Samuel George Morton's Cranial Race Science. *PLoS Biology, 16*(10), e2007008. https://doi.org/10.1371/journal.pbio.2007008
Morning Edition. (2020, June 5). *Minority Workers See Highest Levels of Unemployment from COVID-19 Crisis*. Retrieved September 24, 2020, from www.npr.org/2020/06/05/870227952/minority-workers-see-highest-levels-of-unemployment-from-covid-19-crisis
Nature. (2020). Henrietta Lacks: Science Must Right a Historical Wrong. *Nature Editorial*. www.nature.com/articles/d41586-020-02494-z
Nott, J.C. (1851). Nature and Destiny of the Negro. *DeBow's Review, 10*, 329–332.
Price, G.N., Darity, W.J., & Sharpe, R.V. (2020). Did North Carolina Economically Breed-Out Blacks During Its Historical Eugenic Sterilization Campaign? *American Review of Political Economy, 15*(1), 1–22.
Reston, L. (2017, March 10). *Where Trump Gets His Fuzzy Border Math*. Retrieved September 23, 2020, from https://newrepublic.com/article/140951/trump-gets-fuzzy-border-math
Reverby, S.M. (2011). "Normal Exposure" and Inoculation Syphilis: A PHS "Tuskegee" Doctor in Guatemala, 1946–1948. *The Journal of Policy History, 23*(1), 6–28.
Richards, R.J. (2008). *The Tragic Sense of Life: Ernst Haeckel and the Struggle over Evolutionary Thought*. Chicago: University of Chicago Press.
Richwine, J. (2009). *IQ and Immigration Policy*. Doctoral dissertation, Harvard University. ProQuest Dissertations Publishing. Retrieved August 13, 2020, from https://delong.typepad.com/pdf-1.pdf
Rushton, J.P., & Templer, D.I. (2012). Retracted Article: Do Pigmentation and the Melanocortin System Modulate Aggression and Sexuality in Humans as they do in Other Animals? *Personality and Individual Differences, 53*, 4–8. https://www-sciencedirect-com.uri.idm.oclc.org/science/article/pii/S0191886912000840
Sagan, C. (1997). *The Demon-Haunted World: Science as a Candle in the Dark*. London: Headline Book Publishing.
Saini, A. (2019). *Superior: The Return of Race Science*. Boston, MA: Beacon Press.
Skloot, R. (2010). *The Immortal Life of Henrietta Lacks*. New York: Broadway Paperbacks.
Smith, R. (2014). *Types of Mankind: Polygenism and Scientific Racism in the Nineteenth Century United States Scientific Community*. Electronic Thesis Collection, 105. https://digitalcommons.pittstate.edu/etd/105

Spencer, J. (1995, September 17). Black Scholar Wins Battle of the Minds. *Daily Press*. www.dailypress.com/news/dp-xpm-19950917-1995-09-17-9509170010-story.html

Stapleton, S., Bababekov, Y., Perez, N., Fong, Z., Hashimoto, D., Lillemoe, K., Watkins, M., & Chang, D. (2018). Variation in Amputation Risk for Black Patients: Uncovering Potential Sources of Bias and Opportunities for Intervention. *Journal of the American College of Surgeons*, *226*(4), 641–649.

Treisman, R. (2020, September 16). *Whistleblower Alleges "Medical Neglect", Questionable Hysterectomies of ICE Detainees*. Retrieved September 24, 2020, from www.npr.org/2020/09/16/913398383/whistleblower-alleges-medical-neglect-questionable-hysterectomies-of-ice-detaine

Wilkerson, I. (2020). *Caste: The Origins of Our Discontents*. New York: Random House.

Wood, D. (2020, September 23). *As Pandemic Deaths Add Up, Racial Disparities Persist—And in Some Cases Worsen*. Retrieved September 24, 2020, from www.npr.org/sections/health-shots/2020/09/23/914427907/as-pandemic-deaths-add-up-racial-disparities-persist-and-in-some-cases-worsen

# Part V
## Conclusion

# 11

# Conclusion

*In the fall of 2021, RI State Representative Karen Alzate reached out to me. She was looking for students to speak to a commission studying education and BIPOC people in the state. Immediately, I had two students in mind for this task and they both complied with the request to publicly speak their truths.*

*On Wednesday 12/1/21–3:30 p.m. at the RI Statehouse in Room 101, the poorest high school in RI was represented by Darisel Velez (12th grader) and Ajah Johnson (11th grader). These young ladies spoke eloquently about their personal experiences and the social injustices they witnessed. As representatives of racially minoritized groups, both had important contextual experiences that needed to be heard. During one of the exchanges with the group of legislatures, Darisel said,*

> *For me, I am not a math and science person. I'm the opposite–English, social studies . . . those are my strong suits. But with that science class, like even from the beginning of the year, because I'm taking AP biology with Upegui, he makes sure that we understand that science isn't just about what's going on in that classroom and what the College Board is expecting us to cover in this course. Science is just part of our existence; it's just part of our whole entire livelihood in this world, and I would not say that I absolutely love science, but if I'm being honest, AP biology is probably my favorite class that I'm taking this year because it's not just like, "today we're going to be talking about the nucleus and all of these different organelles." It's us talking about real people. Real people who struggled. Real people who were able to be innovative pioneers. Like in class we talk about people like Dr. Lynn Margulis. We talk about Wilma Rudolph, the Olympic runner.*

DOI: 10.4324/9781003409144-16

> *When we talk in class, it's not just a science environment; it's just a human connection environment, where he sees us, we see him, and it's a mutual respect. He is always telling us not to believe people from outside of our community who may not believe in us, because he believes in us! And we do have so much greatness and so much potential, and to be seen in that room it just makes it, like even though it's science and I'm not a big science person, it just makes it just feels so great to be seen, to be represented and to be able to have someone who really believes in you!*

*Listening to these brilliant young ladies succinctly describing how our science class was empowering them to see their own intrinsic powers, I felt such joy. This can, and should be the purpose of our science education, to explore this wonderful human endeavor that allows us to discover our world and then become full participants in it. Science can be such a powerful democratic tool that values empirical evidence rather than personal prejudice.*

*As Darisel put it, it is our job to ensure that the future stewards of the earth feel "seen." Afterall, the young minds in front of us today, will become the problem solvers that will need to fix all the issues they will inherit. When the science we teach empowers our students we are nurturing them so that they can not only learn, but understand the material, and ultimately become contributors to the body of knowledge and our overall world.*

The word "conclusion" at the top of this chapter takes two meanings here. The first, of course, refers to the fact that this is an ending; the last chapter. But "conclusion" can also refer to reaching a reasoned judgment from inputs such as observations or data. And in this Conclusion we reach some judgments, based upon the contextual data that we have presented in Chapters 5–10, that have resulted the curricular prescriptions that we make in Chapters 2–4.

Even though progress has taken place since the *descarado* hate-filled racism of the colonization and enslavement periods of the 16th to 18th centuries, in the recent past, as we have noted earlier, many of the ideas that a majority of Americans were presumed to espouse were disdainfully dismissed as old-school or "neoliberal."[1] Into this category have fallen ideas like globalism; an open door to immigrants generally, but especially to refugees; an ecumenical approach to all religions; respect for science and legitimate authority; validity of facts; the idea that, in the name of "freedom," individuality trumps any social responsibility; and, of course, the legitimacy of racial diversity and equality. With the election of Donald Trump and the rise of the politics of resentment and division, none of these could be taken for granted in the United States even as aspirational goals, which laid an unanticipated, and surely unwanted, burden on BIPOC Americans. Where in a white-supremacist, monocultural, melanin-challenged society was there a place for them?

The goal of our book, therefore, is to take a small, manageable, but significant step towards remediating some of those insecurities by empowering BIPOC young Americans at a critical time in their educational trajectories. Education is commonly seen as the only real solution to intractable social phenomena like racism; and thus we have proposed an educational step towards solving the problem: we are advocating that while bad (e.g., specious) science has historically oppressed people, good science (e.g., robust) can actually liberate them. "Robust science" is this case is that all humanity is a single species, and that differences in morphotype—skin and eye color; appearance and or color of hair; minor differences in facial features or population-averaged body fat—or any other largely superficial, easily observed quality that might be named, are not predictive of intellect, personality, physical capability, rank along an artificial *Scala Naturæ* (itself an artifact of early Christian theology superimposed upon a misreading of evolutionary history) and, most importantly, human value. Indeed, the superficial features that are used to identify "races" are not positively correlated with any human attribute other than, unsurprisingly, "race." The relationship of genes to complex phenomena such as intelligence and personality was grossly mischaracterized, most egregiously by the overenthusiastic eugenicists of the first half of the 20th century, who thought they could manipulate the genetic basis of intelligence, personality, behavior, and inheritance, when in fact they were lightyears from actually understanding these qualities. As reiterated by Sussman (2014, p. 5),

> Although many people still believe that genes, or a series of genes, directly determine some of our most complex behavioral or cognitive characteristics, the reality is much more complicated. Studies now show that each gene is only a single player in a wondrous intricate drama involving nonadditive interactions of genes, proteins, hormones, food, and life experiences and learning that interact to affect us on different levels of cognitive and behavioral functions. Each gene has an effect on multiple types of behaviors and many behaviors are affected by many genes as well as other factors. The assumption that a single gene is causative can lead to unwarranted conclusions and an overinterpretation of any genuine genetic linkage.

There is now a very strong consensus among scientists, at least, that biological races simply do not exist (Sussman, 2014; and many citations therein). Races would be the operational equivalents of sub-species, and biologically speaking, subspecies can only maintain their independent identities if gene flow among them is impeded by some mechanism; generally, either physical—a mountain range, for example—or behavioral—offset mating times, for example. And, with their proclivity for travel and commerce throughout their ~200,000 years on Earth, humans have continuously interacted on a near-global scale for thousands of years.

The point is exemplified by "Cheddar Man," a 10,000-year old fossil human found in Cheddar Valley, Somerset, England (see Saini, 2019). Here was the oldest Briton in existence; and DNA from the fossils suggested blue eyes, curly hair, paired with skin pigmentation with markers suggesting sub-Saharan African origin. Was this early northern European man dark-skinned? Almost certainly so, perhaps disappointing those who found comfort in imagining themselves to be part of a northern European race of tall, blond, blue-eyed *ubermenschen*.

It is now thought that human group mixing has been so pervasive over the years that Saini (2019) speculates that some of the oldest human lineages known from Africa may have once been light skinned; and groups migrating out of Africa in the deep evolutionary past could even have included light-skinned as well as dark-skinned migrations. Skin color being, as we have seen, but a response to sun, the "race" world today, she reminds us, is not that of the past. She notes:

> *Yet, our modern ideas of race are deeply connected to how we look. Our appearance is shorthand for the stereotypes, a means of slotting people into groups and making judgments about them . . . [The] British public couldn't help but assume that Britons had always looked a certain way, even in the distant past . . . [Cheddar Man] was proof that there couldn't be anything eternal or pure about race because once upon a time, not so very long ago in evolutionary terms, most of the people on Earth didn't look like us. They were already human. They were, however distantly, us. But they looked different.*
> (Saini, 2019, p. 113)

The evolution of human diversity, it seems, is better characterized as a "complex, intermixed trellis, [rather] than a simple candelabra" (Templeton, 1998, quoted in Sussman, 2014, p. 4). Indeed, "race" itself has been a moving target; the races identified in the early 20th century appear hopelessly naïve and outdated today.[2]

The critical *time* in students' educational trajectories is high school. As we discuss in Chapter 1, high school is the right time developmentally for first encountering these ideas, and it is likewise the first time that students will be introduced to a serious thoughtful look at American history. That being the case, should it not reflect events as historians understand them to have actually occurred? We posit that education should aim to promote the common good and should be based on common ground—where teachers can guide students into evidence-based exploration, thoughtful reflection, and meaningful action. We therefore propose classrooms where community is forged through the exploration of science in a welcoming and nurturing environment.

Our recommended approach to all these topics is a historically informed one, because we believe that a clear-eyed understanding of the past is the best way—the only way, in fact—to properly understand, synthesize, and respond to events in the present. This is not radical: it is an obvious and recurrent justification for the study of history. In the case of the United States, its original sin was that the concurrence of freedom and oppression (initially, slavery) was somehow acceptable. Of course, it really didn't work philosophically, with the result that, as Wilkerson (2020) showed, an artificial caste system needed to be imposed. And specious science was the tool that was used to justify the castes.

The result of that was a figurative and literal cost that was—and still is—unsupportable. And to be clear, by this we mean not only the moral cost to our humanity, diminished by the acceptance of the oppression of some people for the benefit of others, but also the monetary cost—the cost of squeezing a subset of our population so that it is not permitted to live as the rest of the society is permitted to live. It is easy to understand that the oppression we have highlighted in this book takes a great toll on not only the oppressed by the oppressor as well (see Freire, 1970/2000); a toll in social welfare; a toll in total societal productivity; a toll in goods and services; a toll in virtually every societal function. One is reminded of the cost to Spanish society in 1492 when all Jews and, shortly thereafter, Muslims were expelled from Spain and, to cite a more recent example, the cost to German society when Jews were expelled from Germany, Austria and, ultimately, occupied lands to the east. What more obviously self-defeating act could either of these two empires have done to themselves?

And yet under the guise of "patriotism," some Americans would prefer to sweep the historical record where it is most easily forgotten, and perpetrate a suite of caricatures about America; its nascence; its exceptionalism; and its moral superiority to the Old World (to recall a recent example from the presidency of George W. Bush). Indeed, even during the review process in the writing of this book, several reviewers asked how we might handle the real possibility that some schools might not adopt our suggested curricular modifications either because of the discomfort that a clear-eyed look at ourselves might engender in some students (and/or their parents and school boards), or because the book has perhaps, more than a whiff of what has popularly but mistakenly been dubbed "Critical Race Theory" and "Woke-ness," both of which have become highly politicized, and anathema, in some circles.

For us, presenting the curricular approaches proposed in this book is an imperative. Rather than lodged between those people who are actively looking for a way to move race relations forward, and those who need to be convinced that what we propose is a worthy endeavor, we hope that our proposal will attract both camps: the first looking right away for the "how to do this" prescription, and the second interested in making a positive contribution to the US's race problem and, seeing our book, becoming convinced that our recommendations are a step in the right direction. Some of the preaching will naturally be to the converted (the first camp), but the book will appeal to them because it gives them a clear prescription for something positive that they can do. The second camp may contain people who remain unconvinced—and uncomfortable looking at the darker side of our history but this for us—and we hope for everybody—is a moral issue: it is no longer possible to sit back and allow "business-as-usual" to transpire. In short, this is our world-view; a consequence of our moral stance—buttressed by the science—and we do not see it as a problem that our book may not be comfortable for all of the people all of the time.

Since 1942, three generations of Americans (including both of us), have daily recited the "Pledge of Allegiance," whose famous ending is, of course:

*One nation, Under God,*[3] *indivisible, with liberty and justice for all.*

Is it not time for the final clause of the Pledge of Allegiance to no longer require disclaimers and omissions, but inevitably to be the truth, the whole truth and nothing but the truth?

*Darisel is completing her first year at Brown University focusing on Political Science. Her first semester she voiced her fears about not finding success. I know well that feeling brought about by the "imposter syndrome." And although she may be uncertain about her capacities, I know she has all the necessary tools to succeed. More importantly, she has successfully utilized those transportable skills she sharpened in her biology class–demonstrating that observation, systems thinking, argumentation, and communication are tools of liberation.*

*Ajah is currently making the challenging decision about where she will attend college after high school graduation. Like Darisel, Ajah will surely use her innate abilities to make our world a better place. Their biology classes allowed them to explore our natural connections to the biota, as well as the unfair socially-constructed paradigms that have limited opportunities for so many.*

*As a tool for student empowerment, education remains the best—and perhaps the only—way to lead our world to a more just society, free from violence, fear, racism, and sorrow. Ideas of the nurturing and empowering nature of education have been around for a long time, but have often found violent opposition from those who benefit from an education system that favors some over others. Oppression in education, then, is the natural extension of social oppression. However, we now have the tools to show that the oppression we see in the world is morally wrong and unsupported by science. Now is the time to create classrooms where our students learn science and simultaneously use the intellectual skills they acquire in so doing to better their world. As the Indian philosopher J. Krishnamurti wrote in his (1982) book,* Questions and Answers, *now is time to rethink the goals of our schools:*

> *"School" comes from the Greek word for leisure—leisure in which to learn, a place where students and teachers can flower, a place where a future generation can be prepared, because schools are meant for that, not just merely to turn out human beings as mechanical, technological instruments—though jobs and careers are necessary—but also flower as human beings, without fear, without confusion, with great integrity.*
>
> (p. 87)

## Notes

1 The Stanford Encyclopedia of Philosophy (2021) defines neoliberalism as "a distinctive political theory. Neoliberalism holds that a society's political and economic institutions should be robustly liberal and capitalist, but supplemented by a constitutionally limited democracy and a modest welfare state." In some tellings, it is a market-centered, near-libertarian approach to markets, free from any external [non-market] constraint (see Oreskes & Conway, 2023).

2 Exemplifying this, one of us (Fastovsky) owns a 1921 map absurdly entitled "The World Problem of Color." The races mapped on it are "White," "Black," "Brown," and "Yellow," with certain mixtures: Central and South America, mixed white and brown; northern Africa, mixed white and black; and eastern Brazil, mixed white, brown, and black.

3 "Under God" was added in 1954 to highlight the distinction between the United States and "atheistic" Communism.

## References Cited

Freire, P. (1970/2000). *Pedagogy of the Oppressed*. New York: Continuum.

Krishnamurti, J. (1982). *Questions and Answers*, p. 107. London: Krishnamurti Foundation.

Oreskes, N., & Conway, E.M. (2023). *The Big Myth: How American Business Taught Us to Loathe Government and Love the Free Market*. New York, NY: Bloomsbury Publishing.

Saini, A. (2019). *Superior: The Return of Race Science*. Boston: Beacon Press.

Stanford Encyclopedia of Philosophy, Neoliberalism. (2021, June 9). Retrieved February 7, 2023, from https://plato.stanford.edu/entries/neoliberalism/

Sussman, R.W. (2014). *The Myth of Race: The Troubling Persistence of an Unscientific Idea*. Cambridge, MA: Harvard University Press.

Templeton, A.R. (1998). Human Races: A Genetic and Evolutionary Perspective. *American Anthropologist, 100*, 632–650.

Wilkerson, I. (2020). *Caste: The Origins of Our Discontents*. New York: Random House.

# Index

For Product Safety Concerns and Information please contact our EU
representative GPSR@taylorandfrancis.com
Taylor & Francis Verlag GmbH, Kaufingerstraße 24, 80331 München, Germany

www.ingramcontent.com/pod-product-compliance
Ingram Content Group UK Ltd.
Pitfield, Milton Keynes, MK11 3LW, UK
UKHW052225190926
475702UK00007B/1208
* 9 7 8 1 0 3 2 5 2 3 8 4 2 *